FARM MACHINERY

STUDENTS ADJUSTING BINDER ATTACHMENTS AFTER THE ATTACHMENTS HAVE BEEN DISMANTLED AND ASSEMBLED

FARM MACHINERY

PRACTICAL HINTS FOR HANDY-MEN

by
J. Brownlee Davidson
and
Leon Wilson Chase

Foreword by Nathaniel Tripp

THE LYONS PRESS

Originally published in 1908 by Orange Judd Company, New York

First Lyons Press edition, 1999

Printed in Canada

10 9 8 7 6 5 4 3 2 1

Library of Congress Cataloging-in-Publication Data

Davidson, J. Brownlee (Jay Brownlee), 1880–1957.
Farm machinery : practical hints for handy men /
by J. Brownlee Davidson
and Leon Wilson Chase ; foreword by Nathaniel Tripp. —
1st Lyons Press ed.
p. cm.
Rev. ed. of: Farm machinery and farm motors. 1908.
ISBN 1-55821-951-X
1. Agricultural machinery. I. Chase, Leon Wilson. II. Davidson, J.
Brownlee (Jay Brownlee), 1880–1957. Farm machinery and farm
motors. III. Title.
S675.D25 1999
631.3—dc21   99-32979
                                                      CIP

FOREWORD

Today we often find them rusting and returning to the earth in the back corner of a field, or behind a weathered, sway-backed barn. Sometimes they appear whole in front of antique dealers, or in part as bar stools. Their original function may be hard to discern; the complexity of design can seem almost comical to unknowing suburbanites. But these relics must not be taken lightly. They are the skeletons of a glorious past, and to know them is to know America.

Our country was born of revolution; political, industrial, and agricultural, each revolution leading in turn to the next. This was profoundly felt on the farm, where most Americans still lived, by the last third of the nineteenth century. After a prolonged period of relative stagnation, American farmers began to be galvanized by the same excitement that was gripping the cities. Cheap, high-quality steel, laid end to end, was opening up the West to settlement, opening up markets in the East. Inventors whose names are still known today, such as Deere and McCormick, applied their enlightened thinking to farm chores, which in some cases had changed little in thousands of years. They shaped the steel to new and complex tasks; these were more than labor-saving devices, these were the means for transformation of the American farmer from slave to master of the land.

This revolution was intellectual as well. In 1862 the Morrill Act set in motion the establishment of land-grant colleges "for the Benefit of Agriculture and Mechanic Arts." This was followed by the creation of agricultural experiment stations in 1887, and the cooperative extension service in 1914.

Agriculture was becoming a science, an industry instead of a means of subsistence, and ideally the "modern" farmer himself was a man of education and enlightened thinking. Nowhere else was the "can do" spirit of American inventiveness more visible than on the farm. Toward the close of the century, great combines drawn by hitches of twenty-four to thirty-six horses were harvesting wheat from one hundred acres a day. These great argosies of the Plains, built of steel and wood and canvas, were as magnificent as the steam locomotives and sailing ships that brought their harvest to the rest of the world. And the rest of the world took notice.

There is elegance, and complex engineering, in the simpler tools as well. A wooden spoked wagon wheel, composed of many precisely fitted parts, constructed of at least two kinds of wood and various grades of steel, is deserving of close study. It combines light weight and sturdiness with an elegance of design, just as the early aeroplanes did, and with similar materials. Its care and maintenance required a certain amount of education and discipline too, and it is all to be found here. More than an historic manual, this book provides an insight into the mind of America at the turn of the century, creative and unfettered by the past. It was also a magical time when technology was still visible and accessible. Repairs could be made, new inventions could be crafted at the home shop and forge, and the farmer, always versatile, truly became a Renaissance man.

Victims of their own success, today's farmers are far fewer and farther between. The technology of improvement is largely invisible too, taking place at the cellular and molecular level. Some complaints heard today about modern agriculture are a curious refrain of fears expressed long ago, that "too much iron would poison the soil," that for every new machine a job was lost, that a revered way of life was changing forever.

Twenty-five years ago, I purchased a hill farm more out of nostalgia for the past and suspicion of the future than understanding of the present. With dreams of farming with horses, I picked up all the equipment I needed for only twice the money required in 1900, but now mine, too, sits abandoned and neglected. When spring comes, bringing with it planting time, birds nest on the hay grapple and the thresher. Wasps live in the seed drill. The sulky plow and mowing machine lie idle while 1950s-era tractors do what little work there is. But I still like to ponder the old machines now and then, feel their castings and forgings, marvel at how anything can be so complex and simple at once. They still provide a harvest of sorts, evoking the proud voices of the men who made and used them.

—Nathaniel Tripp
St. Johnsbury, Vermont, 1999

FARM MACHINERY

PART I

INTRODUCTION

1. One of the requirements for a steady, healthy growth of any people or nation is a bountiful supply of food. The earth can be made to produce in abundance only when the soil is tilled and plants suitable for food are cultivated. As long as the people of the earth roamed about obtaining their subsistence by hunting and fishing, conditions were not favorable for a rapid increase in population or an advance in civilization. Tribes or nations constantly encroached upon each other's rights and were continually at war. History shows that when any nation, isolated so as to be protected from the attacks of other nations, devoted itself to agricultural pursuits, its government at once became more stable and life and property more secure. Protected in this way, a great nation, shut off from the rest of the world by natural means, and located in a fertile country, arose along the banks of the Nile long before any other nation reached prominence. The Gauls became mighty because they devoted themselves to agriculture and obtained in this way a more reliable supply of food. Pliny, the elder, in his writings tells of the fields of Gaul and describes some of the tools used. It has been estimated that there never were more than 400,000 Indians in North America, and they were often in want of food. Compare this number with the present population. The tribes that flourished

and increased in numbers were those who had fields of grain and a definite source of food.

2. Change from hand to machine methods.—When people began to turn their attention to farming they began to devise tools to aid them in their work. Various kinds of hoes, crude plows, sickles, and scythes were invented, but were practically all hand tools. Work with these was necessarily very laborious and slow. The hours of labor in consequence were very long, and the social position of the tiller of the soil was low. He was in every sense of the term "the man with the hoe." He became prematurely old and bent; his lot was anything but enviable.

For more than 3,000 years the farmers of Europe, and in this country until after the Revolutionary War, used the same crude tools and primitive methods as were employed by the Egyptians and the Israelites. In fact, it has been, relatively speaking, only a few years since the change from hand to machine methods took place. In the Twelfth Census Report the following statement is made: "The year 1850 practically marks the close of the period in which the only farm implements and machinery other than the wagon, cart, and cotton gin were those which, for want of a better designation, may be called implements of hand production."

McMaster, in his "History of the People of the United States," says: "The Massachusetts farmer who witnessed the Revolution plowed his land with the wooden bull plow, sowed his grain broadcast, and when it was ripe, cut it with a scythe and threshed it out on his barn floor with a flail." He writes further that the poor whites of Virginia in 1790 lived in log huts "with the chinks stuffed with clay; the walls had no plaster, the windows had no glass, the furniture was such as they themselves had made. Their grain was threshed by driving horses over it in the open field. When they ground it they used a rude pestle and mortar, or placed it in a hollow of a stone and beat it with another."

3. Effects of the change.—At any rate, a great change has taken place and all in little over a half century. This great change from the simplest of tools to the modern, almost perfect implements, has produced a marked effect upon the life of the farmer. He is no longer "the man with the hoe," but a man well trained intellectually.

4. Physical and mental changes.—It is not difficult to realize that a great change for the better has taken place in the physical and mental nature of the farmer. It is vastly easier for a man to sit on a modern harvester, watch the machine, and drive the team, than it is to work all day with bended back, scuffling along, running a cradle. How much easier it is to handle the modern crop, though much larger, with the modern threshing machine, where the bundles are simply thrown into the feeder, than to spend the entire winter beating the grain out with a flail. The farmer can now do his work and still have time to plan his business and to think of improvements.

5. Length of the working day.—One of the marked effects of the change to modern machinery methods has been a shortening of the length of the working day. When the work was done by hand methods, the day during the busy season was from early morn till late at night. Often as much as 16 hours a day were spent in the fields. Now field work seldom exceeds 10 hours a day.

6. Increase in wages.—According to McMaster,* in 1794 "in the States north of Pennsylvania" the wages of

*McMaster: "History of the People of the United States," Vol. II., p. 179.

the common laborer were not to exceed $3 per month,
and "in Vermont good men were employed for £18 per
year." Even as late as 1849, the wages, according to several authorities, did not exceed $120 a year. Under pres-
ent conditions, the farm laborer is able to demand two,
three, and even five times as much. In countries where
hand methods are still practiced, wages are very low.
Men are required to work all day from early morning till
late at night for a few cents. In some of the Asiatic
countries it is said that men work from four in the morning until nine at night for 14 cents. Women receive only
9 or 10 cents and children 7 or 8 cents.

7. **The labor of women.**—Woman, so history relates,
was the first agriculturist. Upon her depended the planting and tending of the various crops. She was required
to help more or less with the farm work as long as the
hand methods remained. Machinery has relieved her of
nearly all field work. Not only this, but many of the
former household duties have been taken away. Spinning
and weaving, soap-making and candle-making, although
formerly household duties, are now turned over to the
factory. Butter and cheese making are gradually becoming the work of the factory rather than that of the home.
Sewing machines, washing machines, cream separators,
and numerous other inventions have come to aid the
housewife with her work.

8. **Percentage of population on farms.**—During the
change from hand to machine methods there was a great
decrease in the percentage of the people of the United
States living upon the farms. It has been estimated that
in 1800 97 per cent of the people were to be found upon
the farms. By 1849 this proportion had decreased to 90
per cent, and according to the Twelfth Census Report it
was only 35.7 per cent.

9. Increase in production.—Notwithstanding this decrease in the per cent of the people upon the farms, there has been, since the introduction of machinery, a great increase in production per capita. In 1800 it is estimated that 5.50 bushels of wheat were produced per capita; in 1850, according to the Division of Statistics of the Department of Agriculture, production had decreased to 4.43 bushels. This was before the effect of harvesting machinery had begun to be felt. People were leaving the farms and the production of wheat per capita was falling off. The limit with hand methods had been reached. Economists were alarmed lest a time should come when the production would not supply the needs of the people. Through the aid of machinery the production increased to 9.16 bushels per capita in 1880, 7.48 bushels in 1890, and 8.66 bushels in 1900. Perhaps this also shows that the maximum production of wheat per capita with present machinery has been reached. The production of corn has also increased, but the increase is not so marked. The production of corn per capita in 1850 was 25.53 bushels; in 1900 it was 34.94 bushels.

10. Cost of production.—Although the cost of farm labor has doubled or trebled, the cost of production has decreased. According to the Thirteenth Annual Report of the Department of Labor, the amount of labor required to produce a bushel of wheat by hand was 3 hours and 3 minutes, and now it is only 9 minutes and 58 seconds. The cost of production, as compiled by Quaintance,* was 20 cents by hand (1829-30) and 10 cents by machinery (1895-96). It is also stated in the Year Book of the Department of Agriculture for 1899 that it formerly required 11 hours of man labor to cut and cure 1 ton of hay. Now

*The Influence of Farm Machinery on Production and Labor. Publications of the American Economic Association, Vol. V., No. 4.

the same work is accomplished in 1 hour and 39 minutes. The cost of the required labor has decreased from 83 1/3 cents to 16 1/4 cents a ton. Not only is it true that machinery has revolutionized the work of making hay, but nearly every phase of farm work has been essentially changed.

11. **Quality of products.**—Machinery has also improved the quality of farm products. Corn and other grains are planted at very nearly the proper time, owing to the fact that machinery methods are so much quicker. By hand methods the crop did not have time to mature. It was necessary to begin the harvest before the grain was ripe, and hence it was shrunken. The grain is obtained now cleaner and purer. It would be difficult at the present time to sell, for bread purposes, grain which had been threshed by the treading of animals over it.

12. **Summary.**—Great changes can be accounted for by the introduction of machine methods for hand methods. For all people this has been beneficial. It has caused the rise of our great nation on the Western Hemisphere. To no class, however, has this change been more beneficial than to the farm worker himself. J. R. Dodge summarized the benefits derived by the farm worker when he wrote: "As to the influence of machinery on farm labor, all intelligent expert observation declares it beneficial. It has relieved the laborer of much drudgery; made his work and his hours of service shorter; stimulated his mental faculties; given an equilibrium of effort to mind and body; made the laborer a more efficient worker, a broader man, and a better citizen."*

Conditions in America have been very favorable for the development of machinery. We have never had an

*American Farm Labor in Rept. of Ind. Com. (1901), Vol. XI., p. 111.

abundance of farm labor. The American inventor has surpassed all others in his ability to devise machines. By this machinery the farmer receives good compensation for his services and is able to compete on foreign markets with cheap labor of other countries.

Lastly, it seems conclusive that an agricultural college course is not complete in which the student does not study much about that which has made his occupation exceptionally desirable. It should be an intensely practical study, for under present conditions success or failure in farming operations depends largely upon the judicious use of farm machinery.

CHAPTER I

DEFINITIONS AND MECHANICAL PRINCIPLES

13. Agricultural engineering is the name given to the agricultural achievements which require for their execution scientific knowledge, mechanical training, and engineering skill.

It has been but quite recently that departments have been organized in agricultural colleges to give instruction in agricultural engineering. The name is not as yet universally adopted, the term farm mechanics or rural engineering being preferred by some. It is hoped that in time "agricultural engineering" will be generally accepted, as it seems to be the broadest and most appropriate term to be given instruction defined as above. Implement manufacturers in Europe have been pleased to call themselves agricultural engineers, and the term is not altogether a new one.

Agricultural engineering embraces such subjects as: (1) farm machinery, (2) farm motors, (3) drainage, (4) irrigation, (5) road construction, (6) rural architecture, (7) blacksmithing, and (8) carpentry.

14. Farm machinery.—Part I. of this treatise, after the present chapter of definitions and mechanical principles and chapters on the transmission of power and the strength of materials, will be a discussion of the construction, adjustment, and operation of farm machinery, and will include the major portion of the implements and machines used in the growing, harvesting, and preparing of farm crops, exclusive of those used in obtaining power. These will be considered in Part II. under the title of

Farm Motors. The following definitions and explanations will prove helpful:

15. A **force** produces or tends to produce or destroy motion. Forces vary in magnitude, and some means must be provided to compare them. Unit force corresponds to unit weight and is the force of gravitation on a definite mass. This unit is arbitrarily chosen and is called the **pound.** The magnitude of all forces, as the draft of an implement, is measured in pounds. Forces also have direction and hence may be represented graphically by a line. For this reason a force is sometimes called a **vector quantity.** Two or more forces acting on a rigid body act as one force called a **resultant.**

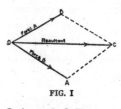

FIG. I

Thus in Fig. 1, O A and O B represent in direction and magnitude two forces acting through the point O. O C is the diagonal of a parallelogram of which O A and O B are sides, and represents the combined action of the forces represented by O A and O B, or is the resultant of these forces. This principle is known as the **parallelogram of forces.**

16. Mechanics is the science which treats of the action of forces upon bodies and the effect which they produce. It treats of the laws which govern the movement and equilibrium of bodies and shows how they may be utilized.

17. Work.—When a force acts through a certain distance or when motion is produced by the action of a force, work is done. Work can therefore be defined as the product of force into distance. Work can be defined in another way as being proportional to the distance through which the force acts, and also to the magnitude of the force.

18. Unit of work.—It has been stated that the unit of force is the pound. The unit of distance is the foot. The unit of work is unit force acting through unit distance and is named the **foot-pound.** A foot-pound is then the amount of work performed in raising a mass weighing 1 pound 1 foot. It is to be noted that the amount of work done in raising 1 pound through 10 feet is the same as raising 10 pounds through 1 foot. It is to be noted further that, in considering the amount of work, time is not taken into account. It is the same regardless of whether 1 minute or many times 1 minute was used in performing the operation. The **horse-power hour** is another unit of work commonly used and will be understood after power has been defined.

19. Power is the rate of work. To obtain the power received from any source the number of foot-pounds of work done in a given time must be determined. The unit of power commonly used is the horse power.

20. A **horse power** is work at the rate of 33,000 foot-pounds a minute, or 550 pounds a second. That is, if a weight of 33,000 pounds be raised through 1 foot in 1 minute, one horse power of work is being done. This unit was arbitrarily chosen by early steam engine manufacturers to compare their engines with the power of a horse.

If a horse is walking 2.5 miles an hour and exerting a steady pull on his traces of 150 pounds, the effective energy which he develops is:

$$\frac{150 \times 5280 \times 2.5}{60 \times 33000} = 1 \text{ H. P.}$$

21. A **machine** is a device for applying work. By it motion and forces are modified so as to be used to greater advantage. A machine is not a source of work. In fact, the amount of work imparted to a machine always exceeds the amount received from it. Some work is used in overcoming the friction of the machine. The ratio between the amount of work received from a machine and the amount put into it is called the **efficiency of the machine.**

22. Simple machines are the elements to which all machinery may be reduced. A machine like a harvester, with systems of sprockets, gears, and cranks, consists only of modifications of the elements of machines. These elements are six in number and are called (1) the lever, (2) the wheel and axle, (3) the inclined plane, (4) the screw, (5) the wedge, and (6) the pulley. These six may be conceived to be reduced to only two—the lever and the inclined plane.

23. The law of mechanics holds that the power multiplied by the distance through which it moves is equal to the weight multiplied by the distance through which it moves. Thus, a power of 1 pound moving 10 feet equals 10 pounds moving 1 foot. This is true in theory, but in practice a certain amount must be added to overcome friction.

24. The lever, the simplest of all machines, is a bar or rigid arm turning about a pivot called the fulcrum. The object to be moved is commonly designated as the weight, and the arm on which it is placed is called the weight arm. The force used is designated as the power, and the arm on which it acts is called the power arm. Levers are divided into three classes; for an explanation of the classes refer to any text on physics.* The law of mechanics may be applied to all levers in this manner. The power multiplied by the power arm equals the weight multiplied by the weight arm.

*"General Physics." By C. S. Hastings and F. E. Beach and others.

If P = Power, Pa = Power arm, W = Weight, and Wa = Weight arm, P × Pa = W × Wa.

If three of these quantities are known, the other is easily calculated. The arm or leverage is always the perpendicular distance between the direction of the force and the fulcrum.

25. The two-horse evener or doubletree.—The two-horse evener is a lever of the second class where the clevis pin for the whiffletree at one end acts as the fulcrum for

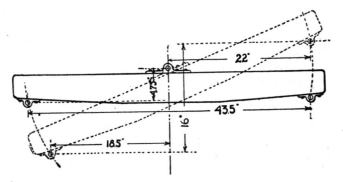

FIG. 2—WAGON EVENER IN OUTLINE. SHOWING THE ADVANTAGE THE LEADING HORSE HAS WHEN THE CLEVIS HOLES ARE NOT A STRAIGHT LINE

the power applied by the horse at the other end. The weight is the load at the middle. If the three holes for the attachment of each horse and the load be in a straight line and the arms be of equal length, each horse pulls an equal share of the load even if the evener is not at right angles with the line of draft. But more often the end holes in the evener are placed in a line behind the hole for the center clevis pin. Then if one horse permits his end of the evener to recede, he will have the larger portion of the load to pull because his lever arm has been shortened more than the lever arm of the other horse. The author's attention has been called to a wagon doubletree in which the center and end holes for clevis pins are made ' by iron clips riveted to the front and back sides of the wood. The center hole was thus placed 4¾ inches out of the line of the end holes. This evener is shown in outline in Fig. 2.

By calculation it was found that if one horse was 8 inches in the advance of the other, the rear horse would pull 8.64 per cent more than the first, or 4.32 per cent more of the total load. If this difference was 16 inches, the rear horse would pull 19 per cent more than the first, or 8 per cent more of the total load.

26. Eveners.—When several horses are hitched to a machine as one team, a system of levers is used to divide the load proportionately. The law of mechanics applies in all cases, noting that the lever arm is the perpendicular distance between the direction of the force and the fulcrum or pivot. In general, it may be said that there is nothing to be gained by a complicated evener. If there is a flexible connection and an equal division of the draft, the simple evener is as good as the complicated or so-called "patent" evener. The line of draft cannot be offset without a force acting across it. This is accomplished with a tongue truck, which seems to be the logical method.

Fig. 3 illustrates some good types of eveners.

27. Giving one horse the advantage.—It often occurs in working young animals or horses of different weights that it is desired to give one the advantage in the share of work done. This is accomplished by making one evener arm longer than the other, giving the horse which is to have the advantage the longer arm. This may be done by setting out his clevis, setting in the clevis of the other horse, or placing the center clevis out toward the other horse. The correct division of the load between horses of different sizes is not definitely known, but it is

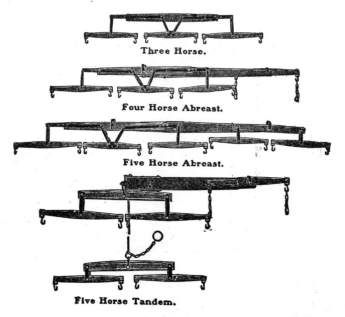

Three Horse.

Four Horse Abreast.

Five Horse Abreast.

Five Horse Tandem.

FIG. 3—GOOD TYPES OF EVENERS WHICH WILL DIVIDE EQUALLY THE DRAFT

thought that the division should be made in about the same proportion as each horse's weight is of their combined weight.

28. Inclined plane.—The tread power is an example of the utilization of the inclined plane, in which the plane is an endless apron whose motion is transferred to a shaft. The tread power is illustrated in Part II., Farm Motors.

29. The screw is a combination of the inclined plane

and the lever, where the inclined plane is wrapped around a cylinder and engages a nut. The **pitch** of a screw is the distance between a point on one thread to a like point on the next, or, in other words, it is 1 inch divided by the number of threads to the inch. Thus, 8 threads to 1 inch is 1/8 pitch, 24 threads 1/24 pitch. There is a great gain of power in the screw because the load is moved a short distance compared with the power. A single-pitch thread advances along the length of the screw once the pitch at each turn; a double pitch advances twice the pitch. The part of a bolt containing a screw thread on the inside is spoken of as a **nut.** The name burr is often given to the nut, but **burr** applies more particularly to washers for rivets. The tool used in making the thread in a nut is called a **tap,** and the one for making outside threads a **die.**

FIG. 4—SIMPLE PUL-LEY, WHICH ONLY CHANGES THE DI-RECTION OF A FORCE

30. A **pulley** consists primarily of a grooved wheel and axle over which runs a cord.

A simple pulley changes only the direction of the force. By a combination of pulleys the power may be increased indefinitely. The wheel which carries the rope is called a **sheave,** the covering and axle for the sheave the **block,** and the whole a **pulley.** A combination of blocks and ropes is called a **tackle.** With the common tackle block, the power is multiplied by the number of strands of rope less one.

The mechanical advantage may be obtained in another

way, as it is equal to the number of strands supporting the weight. This will agree with the former method when the power is acting downward. If the power is acting upward instead of downward, the power strand would be supporting the weight, and so should not be deducted from the total number to obtain the mechanical advantage.

Fig. 5 illustrates a tackle which has six strands, but

FIG. 5—A TACKLE. A FORCE MAY BE MULTIPLIED MANY TIMES BY A TACKLE OF THIS KIND

FIG. 6 — DIFFEREN. TIAL TACKLE BY WHICH HEAVY WEIGHTS MAY BE RAISED

only five are supporting the weight, so the mechanical advantage in this case is five. If the weight be 1,000 pounds, as marked, a force of 200 pounds besides a force sufficient to overcome friction will be needed to raise the weight. This tackle has a special designed sheave which, when the free rope end is carried to one side and let out slightly, the rope is wedged in a special groove and the weight held firmly in place.

The **differential pulley** shown in Fig. 6 is a very powerful device for raising heavy weights and is very simple. The principle involved is that the upper sheaves are of different diameters, fastened rigidly together and engaging the chain in such a manner as to prevent it from slipping over them. Thus, as the sheaves are rotated, one of the strands of chains carrying the load is taken up slightly faster than the other is let out, shortening their combined length and raising the load.

31. Dynamometers* are instruments used in determining the force transmitted to or from a machine or imple-

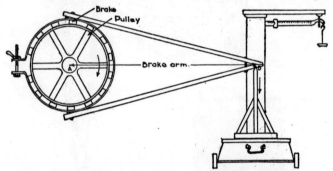

FIG. 7—PRONY BRAKE: ONE FORM OF ABSORPTION DYNAMOMETER

ment. They are, therefore, very important instruments for the study and testing of machinery. Having determined with this instrument the force, it is an easy matter to calculate the power.

32. Absorption dynamometers are those which absorb the power in measuring the force transmitted. The **Prony brake** as illustrated in Fig. 7 is the common device used

*For additional literature on the measurement of power see "Experimental Engineering," by R. C. Carpenter.

in measuring the output of motors. The force transmitted is measured by a pair of platform scales or a spring balance. The distance through which this force acts in 1 minute is calculated from the number of revolutions of the rotating shaft per minute and the distance through which the force would travel in one revolution if released. The revolutions of the shaft are obtained by means of a speed indicator, a type of which is illustrated in Fig. 8.

FIG. 8—SPEED INDICATOR: AN INSTRUMENT FOR DETERMINING THE SPEED

If π = ratio between diameter of circle and the circumference = 3.1416,

  a = length of brake arm in feet,

  G = net brake load (weight on scale less weight of brake on scale),

  n = revolutions a minute,

$$H. P. = \frac{2 \pi G a n}{33000}$$

Dynamometers which do not absorb the power are called **transmission dynamometers**.

33. Traction dynamometers.—Dynamometers used in connection with farm machinery to determine the draft of implements are called **traction dynamometers**. They are instruments on the principle of a pair of scales placed between an implement and the horses or engine. They indicate the number of pounds of draft or pull required to move the implement. The traction dynamometer is a transmission dynamometer. The power is not all used up in the measuring, but transmitted to the implement or machine where the work is being done.

The operation of the traction dynamometer is the same as that of a heavy spring balance. The spring may be a coil, flat or elliptical, or an oil or water piston may be used in place of the spring and the pull determined by the pressure produced.

34. Direct-reading dynamometers.—The more simple types of dynamometers have a convenient scale and a needle which indicates the pull in pounds. A second needle is usually provided which shows the maximum pull which has been reached during the test. A dynamometer of this kind is illustrated in Fig. 9. This has

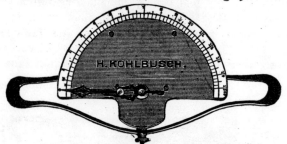

FIG. 9—DIRECT-READING DYNAMOMETER

elliptical springs and a dial upon which the draft is registered. It is difficult to obtain accurate readings from a dynamometer of this sort on account of vibration caused by the change of draft due to rough ground or the unsteady motion of the horse.

35. Self-recording dynamometers.—A recording dynamometer records by a pen or pencil line the draft. A strip of paper is passed under the needle carrying the pen point, whose position is determined by the pull. The height of the pen line above a base line of no load is proportional to the pull in pounds. A diagram obtained in this way is shown in Fig. 10. Often the paper is ruled to scale so that direct readings may be made from the paper. Methods of rotating the reel or spool vary in different makes. Some German dynamometers rotate the

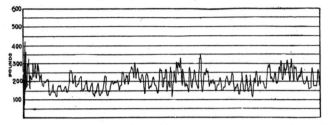

FIG. 10—A RECORD OF DRAFT OBTAINED BY A RECORDING DYNAMOMETER

reel by a wheel which runs along on the ground and is connected to the reel by a flexible shaft, as in Fig. 11. This method is very satisfactory, except that the wheel is

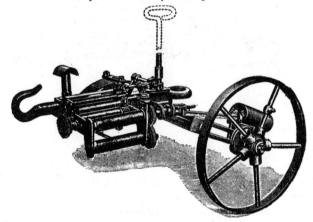

FIG. 11—A GERMAN RECORDING DYNAMOMETER WHICH HAS THE REEL
DRIVEN BY A WHEEL TRAVELING ON GROUND

often in the way. Distances along the paper are in this case proportional to the distance passed over by the implement.

Another method is to rotate the reel by clock-work. Then distances along the paper are proportional to time.

FIG. 12—GIDDINGS RECORDING DYNAMOMETER, WHICH HAS THE REEL DRIVEN BY CLOCK-WORK

If the velocity be uniform, the distances are approximately proportional to the distance passed over as before. When the distances along the paper are proportional to the ground passed over, the amount of work may be obtained easily. The Giddings dynamometer, as illustrated in Fig. 12, is made in this way. It also has elliptical springs.

Still another method is made use of in another type of dynamometer, in which the in-and-out movement of the pull head is made to rotate the reel. This method is not

FIG. 13—THE PLANIMETER USED TO FIND THE AVERAGE DRAFT FROM DYNAMOMETER RECORDS. THERE ARE SEVERAL TYPES OF THIS INSTRUMENT

so satisfactory because distances along the paper are not proportional to anything. If the draft remains constant, there is no rotation of the reel at all. Various devices are provided dynamometers to add the draft for stated distances, and in this way obtain the work done. A tape line 100 feet long is sometimes used to rotate the reel of the dynamometer.

To obtain the mean draft a line is drawn through the graph of the pen point, eliminating the sharp points. Then the diagram may be divided into any number of equal parts and the sum of the draft at the center of these divisions divided by the number of divisions. The quotient will be the mean draft.

An instrument called the **planimeter** (Fig. 13) will de-

termine the area of the diagram when the point is passed
around it. To obtain the mean height and the average
draft it is only necessary to divide the area of the diagram
by its length. This can only be done when distances
along the paper are proportional to the distance passed
over by the implement.

36. Steam and gas engine indicators.—The indicator,
although not used much in connection with farm engines,

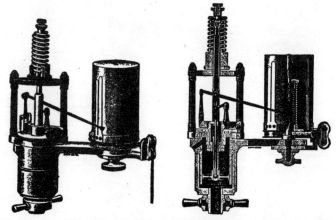

FIG. 14—THE STEAM OR GAS ENGINE INDICATOR. AN INSTRUMENT USED
TO OBTAIN A RECORD OF THE PRESSURE IN THE ENGINE CYLINDER
AT VARIOUS POINTS OF THE STROKE

should be mentioned at this point under a discussion of
the methods of measuring work.

Fig. 14 illustrates a steam engine indicator complete,
and also a section of it showing the mechanism inside. In
brief, the indicator consists in a drum, upon which a paper
card is mounted to receive the record or diagram, and a
cylinder carefully fitted with a piston upon which the
pressures of the steam or gases from the engine cylinder
act. The drum by a mechanism called a **reducing motion**
is given a motion corresponding to that of the engine

piston, and the pressure of the gases from the engine cylinder acting on the piston of the indicator compresses a calibrated spring above. The amount of pressure is recorded with a pencil point by a suitable mechanism on the paper card. Thus if a diagram is obtained from an engine at work, it not only permits a study of the engine in regard to the action of valves, igniter, etc., but also enables the amount of work performed in the engine cylinder to be calculated.

FIG. 15—AN ACTUAL INDICATOR DIAGRAM OBTAINED FROM A GAS ENGINE, WITH THE SCALE OF THE SPRING APPENDED

Fig. 15 shows an actual diagram taken from a gas engine. As the pressure varies throughout the stroke, an instrument like the planimeter of Fig. 13 must be used to average the pressure for the entire working stroke of the piston, and subtract the pressure required in the preliminary and exhaust strokes. This average pressure is called the mean effective pressure (M.E.P.). Knowing the distance the engine piston travels a minute doing work, the area of the surface on which the pressure acts, and the mean effective pressure, it is possible to calculate the rate of work or the horse power. The horse power obtained in this way is called the indicated horse power (I.H.P.), and differs from the brake horse power (B.H.P.) by the power required to overcome friction in the engine. The ratio of the brake horse power to the indicated horse power is called the mechanical efficiency of the engine.

If P = Mean effective pressure,
    L = Length of stroke in feet,
    A = Area of piston in square inches,
    N = Number of working strokes a minute,

$$\text{I. H. P.} = \frac{\text{PLAN}}{33,000}$$

It is to be noted that in double-acting engines the faces of the piston on which the pressure in the engine cylinder acts differ by the area of the cross section of the piston rod. It is customary to calculate the indicated horse power for each end of the cylinder, and take the sum for the indicated horse power of the engine.

37. Heat.—Work, as measured by the foot-pound, is mechanical energy or the energy of motion. Energy is defined as the power to produce a change of any kind and manifests itself in many forms. It may be transformed from one form to another without affecting the whole amount. Heat represents one form of energy, and it is the purpose of all heat engines to transform this heat energy into mechanical energy. Like work, heat may be measured. The unit used for this purpose is the British thermal unit.

The **British thermal unit** (B.T.U.) is the amount of heat required to raise the temperature of 1 pound of water 1° F. To make the unit more specific, the change of temperature is usually specified as being between 62° and 63° F. The work equivalent of the British thermal unit is sometimes called the Joule (J) and is equal to 778 foot-pounds of work.

Thermal efficiency is a term used in connection with heat engines to represent the ratio between the amount of energy received from the engine in the form of work and the amount given to it in the form of heat. The thermal efficiency of a steam engine seldom exceeds 15 per cent and of a gas engine 30 per cent.

38. Electrical energy.—By means of a dynamo, mechanical energy may be converted into electrical energy or the energy of an electric current. An electric current may be likened to the flow of water through a pipe in that it has pressure and volume. In the water pipe the pressure is measured in pounds to the square inch, and the volume by the area of the cross section of the pipe. With an electric current the pressure is measured in **volts** and the volume or amount of current in **amperes.** Thus a current may have a pressure or voltage of 110 volts and a volume or amperage of 7 amperes. The product of volts into amperes gives **watts.** An electrical current of 746 watts is equal to one horse power. Electric energy is bought and sold by the watt-hour, or the larger unit, the kilowatt-hour, which is 1,000 watt-hours.

CHAPTER II

TRANSMISSION OF POWER

It is the function of all machines to receive energy from some source and distribute it to the various parts where it will be converted into useful work. This chapter will treat of the devices used in the transmission and distribution of power and the loss of power during transmission.

39. Belting.—Belting is one of the oldest and most common devices used for the transmission of power from one rotating shaft to another. The transmission depends upon the friction between the belt and the pulley face; that is, the belt clings to the pulley face and causes it to rotate as the belt travels around it. The sides of a belt, when connecting two pulleys and transmitting power, are under unequal tension. The **effectual tension** or actual force transmitted is the difference between the tensions on each side. The effectual tension multiplied by the velocity of the belt in feet a minute will give the foot-pounds of work transmitted a minute. Thus the power varies directly with effectual tension and the velocity of the belt.

40. Horse power of a leather belt.—It is possible to make up a formula with the above quantities to be used in the calculation of the power of a belt or the size required to transmit a certain power. The following is a common rule for single-ply belting, which assumes an effectual tension of 33 pounds an inch of width:

H. P. = Horse power,
   v = Velocity in feet a minute,
   w = Width of belt in inches.

$$\text{H. P.} = \frac{vw}{1000}$$

The quantity v may be calculated from the number of revolutions a minute and the diameter of the driving pulley. The velocity of belts rarely exceeds 4,500 feet a minute. The highest efficiency of belt transmission is obtained from belting when there is no slipping and little stretching, and when the tension on the belt does not create an undue pressure on the bearings.

41. Leather belting.—Good leather belting will last longer than any other when protected from heat and moisture. A good belt should last for 10 to 15 years of continuous service. Best results are obtained when the hair or grain side of the leather is run next to the pulley. When the belt is put on the opposite way, the grain side, which is firmer and has the greater part of the strength of the belt, is apt to become cracked and the strength of the belt much reduced.

42. Care of leather belts.—Belts should be occasionally cleaned and oiled to keep them soft and pliable. There are good dressings upon the market, and others that are certainly injurious. Neatsfoot oil is a very satisfactory dressing. Mineral oils are not very satisfactory, as a rule. Rosin is considered injurious, and it is doubtful if it is necessary to use it on a belt in good condition. With horizontal belts it is desirable to have the under side the driving side, for then the sag of the slack side causes more of the belt to come in contact with the pulleys and will prevent slippage to some extent.

43. Rubber belting.—Good rubber belting is of perfect uniformity in width and thickness and will resist a greater degree of heat and cold than leather. It is especially well adapted to wet places and where it will be exposed to the action of steam. Rubber belting, which clings well to the pulley, is less apt to slip and may be called upon to do very heavy service. Although not as durable as leather, it is quite strong, but offers a little difficulty in the making of splices. Rubber belting is made from two-ply to eight-ply in thickness. A four-ply belt is considered the equal of a single-ply leather belt in the transmission of power. All oil and grease must be kept away from rubber belting.

44. Canvas belting is used extensively for the transmission of power supplied by portable and traction engines. It is very strong and durable, and is especially well adapted to withstand hard service. When used in the field it is usually made into endless belts. It has one characteristic which bars its extended use between pulleys at a fixed distance, and that is its stretching and contracting, due to moisture changes. Canvas belting, like rubber belting, is made in various thicknesses from two-ply up. A four-ply belt is usually considered the equal of a single leather belt.

45. Length of belts.—Length of belts is usually determined after the pulleys are in place by wrapping a tape line around the pulleys. When this cannot be done conveniently, the following approximate rule taken from Kent's Mechanical Engineer's Pocketbook may be used: "Add the diameter of the two pulleys, divide by two, and multiply the quotient by $3\frac{1}{4}$, and add the product to twice the distance between the centers of the shafts."

46. Lacing of belts.—Lacing with a rawhide thong is the common method used in connecting the ends of a belt. A laced belt should run noiselessly over the pulleys and should be as pliable as any part of the belt. The holes should be at least five-eighths inches from the edge and should be placed directly opposite. An oval punch is the best, making the long diameter of the hole parallel with the belt. With narrow belts only a single row of holes need be punched, but with wide belts it is necessary to punch a double row of holes.

By oiling or wetting the end of the lace and then burning to a crisp with a match the lacing may be performed more easily. Begin lacing at the center of the belt and never cross or twist the lace or have more than two thicknesses of lace on the pulley side of the belt. In lacing canvas belts, the holes should be made with a belt awl. When the lacing is finished the lace may be pulled through a small extra hole and then cut so as to catch over the edge. By this method, tying of the lace is avoided. Fig. 16 illustrates four good methods of lacing a belt with a thong.

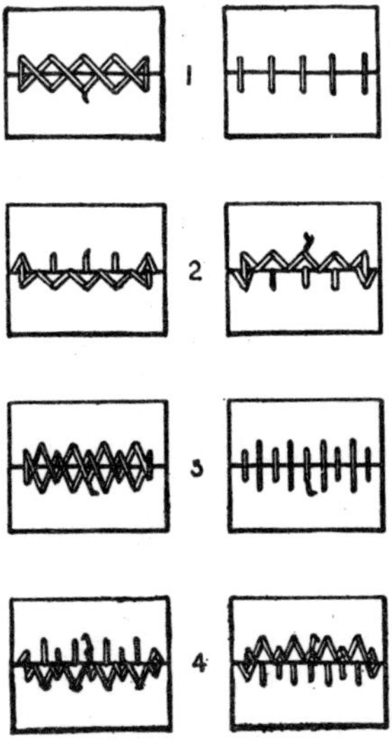

FIG. 16—FOUR GOOD STYLES OF BELT LACING

1 shows a method of lacing a belt with a single row of holes.

2 shows a light hinge lace for a belt to run around an idler.

3 shows a double row lace.

4 shows a heavy hinge lace.

47. Wire belt lacing makes a very good splice. The splice when properly made is smooth and well adapted for leather and canvas belting. When this lacing is used, the holes should be made with a small punch, the thickness of the belt from the edge and twice the thickness apart. The lacing should not be crossed on the pulley side of the belt.

48. **Pulleys.**—Pulleys are made of wood, cast iron, and steel. They are also constructed solid or in one piece and divided into halves. It is best to have a large cast pulley divided, as the large solid pulley is often weakened by contraction in cooling after being cast. For most purposes the iron pulley is the most satisfactory, as it is

neat and durable. Belts do not cling to iron pulleys well, and hence they are often covered with leather to increase their driving power. Often the driving power is increased one-fourth in this way.

Pulleys are crowned or have an oval face in order to keep the belt in the center. The tendency of the belt is to run to the highest point, as shown in Fig. 17. The pulley that imparts motion to the belt is called the driver and the one that receives its motion

FIG. 17—SHOWING THE EFFECT OF CROWN ON PULLEYS

from the belt the driven.

49. **Rules for calculating speed of pulleys.**—Case I. The diameters of the driver and driven and the revolutions per minute of the driver being given, to find the number of revolutions per minute of the driven. Rule: Multiply the diameter of the driver by its r.p.m. and divide the product by the diameter of the driven; the quotient will be the r.p.m. of the driven.

Case II. The diameter and the revolutions per minute of the driver being given, to find the diameter of the driven that shall make any given number of revolutions per minute. Rule: Multiply the diameter of the driver by its r.p.m. and divide the product by the r.p.m. of the driven; the quotient will be its diameter.

Case III. To ascertain the size of the driver. Rule: Multiply the diameter of the driven by the r.p.m. desired that it should make and divide the product by the revolutions of the driver; the quotient will be the size of the driver.

No allowance is made in the above rules for slip.

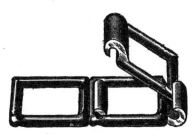

FIG. 18—MALLEABLE LINK BELTING OF ROTATING SHAFTS

50. Link belting.—A common means of distributing power to various parts of a machine is by link belting. Chain link belting is adapted to almost all purposes except high speed. Two kinds of link belting are now found in general use. One style is made of malleable iron links (Fig. 18) and the other crimped steel (Fig. 19). In regard to the desirability of each, data is not at hand. However, it is stated that the steel links wear longer, but

FIG. 19—STEEL LINK BELTING

cause the sprockets to wear faster. If this be true, the steel belting should be used on large sprockets and the malleable confined to the smaller sprockets.

51. Rope transmission often has many advantages over belt transmission in that the first cost of installation is less, less power is lost by slippage, and the direction of transmission may be easily changed. Transmission ropes are made of hemp, manila, and cotton. Cotton rope is not as strong as the others, but is much more durable, especially when run over small pulleys or sheaves. The groove of the pulley or sheaves should be of such a size and shape as to cause the rope to wedge into it, thus permitting the effective tension of rope to be increased to its working strength.

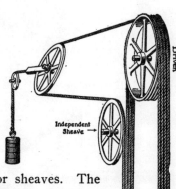

Fig. 20 illustrates a rope transmission system. Transmission ropes, to insure the highest efficiency in respect to the amount of power transmitted and the durability of the ropes, should have a velocity of from 3,000 to 4,000 feet a minute. To lubricate the surface of the rope and prevent it from fraying, a mixture of beeswax and graphite is good.

52. Wire rope or cable transmission.— For transmission of power to a distance and between buildings, wire rope has many advantages. If the distance of transmission be over 500 feet, relay stations with idler pulleys should be installed to carry the rope. Pulleys or sheaves for wire rope should not have grooves into which the rope may wedge, as this is very detrimental

FIG. 20—TRANSMISSION OF POWER BY ROPES

to the durability of the rope. The sheaves for wire rope should have grooves filled with rubber, wood, or other material to give greater adhesion.

Fig. 21 illustrates how a wire rope may be used to transmit power between buildings. For tables useful in

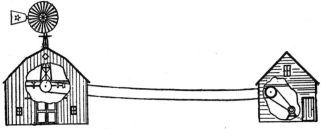

FIG. 21—TRANSMISSION OF POWER BY A WIRE ROPE

determining the size of rope required for a rope transmission, see any engineering handbook. They require too much space to be included in this work.*

53. Rope splice.—To splice a rope the ends should be

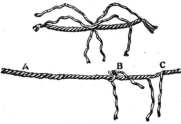

FIG. 22—METHOD OF SPLICING A ROPE

cut off square and the strands unbraided for not less than 2½ feet and crotched together as shown at 1 in Fig. 22. After the strands of one end are placed between the strands of the other, untwist one strand as at C and

*"Mechanical Engineers' Pocket Book." By William Kent.

wind the corresponding strand of the other rope end into
its place until about 9 to 12 inches remain. After this is
done, the strand should be looped under the other, forming the knot shown at B, with the strand following the
same direction as the other strands of the rope. Another
strand is now unwound in the opposite direction and the
same kind of knot formed. The long ends of the unwound strands are cut to the same length as the short
ones, and the short ends woven into the rope by passing
over the adjacent strand and under the next, and so on.
This is continued until the end of the strand is completely woven into the rope. The same operation is fol-

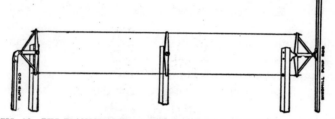

FIG. 23—THE TRANSMISSION OF THE POWER OF A WINDMILL TO A PUMP
AT A DISTANCE BY MEANS OF TRIANGLES AND WIRES

lowed with all of the strands until a smooth splice is obtained. The above directions apply well for splicing
ropes used with haying machinery. The same method
may be used with transmission rope, although with the
latter the splice is often made much longer.

54. Triangles.—A very handy method of transmitting
the power of a windmill to a pump at a distance is by
means of triangles, as illustrated in Fig. 23. These triangles are attached to each other by common wire, and,
if the distance is great, stations with rocker arms are
provided to carry the wires. When triangles are used to
connect a windmill to a pump the wires are often crossed

in order that the up stroke of the pump will be made with the up stroke of the windmill.

55. Gearing.—Spur gears are wheels with the teeth or cogs ranged around the outer or inner surface of the rim in the direction of radii from the center, and their action is that of two cylinders rolling together. To transmit uniform motion, each tooth must conform to a definite profile designed for that particular gear or set of gear wheels. The two curves to which this profile may be constructed are the involute and the cycloid. Gear wheels must remain at a fixed distance from each other, or the teeth will not mesh properly.

Fig. 24 illustrates some of the common terms used in connection with gear wheels. Bevel gears have teeth

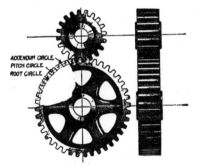

ADDENDUM CIRCLE
PITCH CIRCLE
ROOT CIRCLE

FIG. 24—SPUR GEARING

similar to spur gears, and their action is like that of two cones rolling together.

The teeth of gear wheels are cast or machine cut. Most of the gear wheels found on agricultural machines have the teeth simply cast, as this is the cheaper method of construction. Where smoothness of running is desired, the teeth are machined in, and the form of each tooth is more perfect, insuring smoother action. The cream separator has machine-cut gears. Very large gear wheels have each tooth inserted in a groove in the gear wheel rim. Such a tooth is called a **cog**; hence the term cog is often applied to all forms of the gear tooth. Cogs may be made of metal or wood.

Like pulleys, gear wheels are spoken of as the driver and the driven. To find the speed ratio of gear wheels, the following rule may be used:

Rule: Revolution of driver per minute, multiplied by the number of teeth in driver, equals the revolution of the driven per minute, multiplied by the number of teeth in driven.

56. Shafting.—Where several machines are to be operated from one power unit, it is necessary to provide shafting on which pulleys are placed. Shafting should be supported by a hanger at least every 8 feet, and the pulleys placed as near as possible to the hangers. Thurston gives the following formula for cold-rolled iron shafting:

$$\text{H. P.} = \frac{d^3 R}{55}$$

when H.P. is the horse power transmitted, d is the diameter of shaft in inches, R the revolutions per minute. Steel shafting will transmit somewhat more power than iron, and some difference may be made for the way the power is taken from the shaft; but the above rule is considered a safe average.

57. Friction.—It has been stated that a machine will not deliver as much energy as it receives because a certain amount must be used to overcome friction. **Friction** is the resistance met with when one surface slides over another. Since machines are made of moving parts, friction must be encountered continually. In the majority of cases it is desired to keep friction to a minimum, but in others it is required. In the case of transmission of power by belting it is absolutely necessary.

58. Coefficient of friction is the ratio between the force tending to bring two surfaces into close contact and the force required to slide the surfaces over each other. This force is always greater at the moment sliding begins. Hence it is said that **friction of rest** is greater than sliding friction.

The following table of coefficients of friction is given to show the effect of lubrication (Enc. Brit.):

Wood on wood, dry	0.25 to 0.5
" " " soaped	0.2
Metals on oak, dry	0.25 to 0.6
" " " wet	0.24 to 0.26
" " " soaped	0.2
" " metal, dry	0.15 to 0.2
" " " wet	0.3
Smooth surfaces occasionally lubricated	0.07 to 0.08
" " thoroughly "	0.03 to 0.036

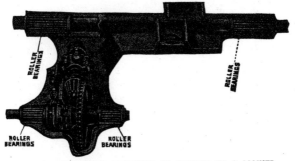

FIG. 25—ROLLER BEARINGS AS APPLIED TO A MOWER

59. Rolling friction.—When a body is rolled over a surface a certain amount of resistance is offered. This resistance is termed rolling friction. Rolling friction is due to a slight compression or indentation of the surfaces under the load, hence is much less with hard surfaces than with soft. Rolling friction is that met with in ball and roller bearings, and is much less than sliding friction. Roller bearings reduce friction greatly. Ball bearings may be used advantageously when end thrust is to be overcome or where they can be used in pairs. They are not suitable for carrying heavy loads.

60. Lubrication.—The object of lubrication is to reduce friction to a minimum. A small quantity of oil is placed in a box and a thin film adheres both to the surface of the journal and also to the bearing, so in reality the friction takes place between liquid surfaces. The lubricant also fills the unevenness of the surfaces, so that there is no interlocking of the particles that compose them. Friction with a lubricant varies greatly with the quality of lubricant and the temperature.

61. Choice of lubricant.—For heavy pressures the lubricant should be thick so as to resist being squeezed out under the load, while for light pressures thin oil should be used so that its viscosity will not add to the friction. Thus, for a wagon, heavy grease should be used, while for a cream separator of high speed a thin oil is necessary. Temperature must also be taken into account in choosing a lubricant.

Solid substances in a finely divided state, such as mica and graphite, are used to reduce friction. The practice seems to be a very good one. This is especially true with graphite in bearings that can be oiled only occasionally, as the bearings of a windmill.

62. Bearings should be of sufficient size that the lubricant will not be squeezed out from between the journal and the bearing. In the design of machinery a certain pressure limit must not be exceeded. It is better to have the journal and bearing made out of different materials, as the friction in this case is less and there is a less tendency for the surfaces to abrade. Brass, bronze, and babbit are used for bearings with a steel journal. It is highly essential that the bearing be kept free from all dirt and grit. Occasionally it is better to let some minor bearings go entirely without lubrication, for the oil only

FIG. 26—A SELF-OILING AND SELF-ALIGNING BEARING. OFTEN THE OIL RESERVOIR BELOW THE RINGS IS ENLARGED AND THE WICK DISPENSED WITH

causes the gathering of grit and sand to grind out the bearing.

63. Heating of boxes may be due to (1) insufficient lubrication, (2) dirt or grit, (3) the cap may be screwed down too tight, (4) the box may be out of line and the shaft may bind, (5) the collar or the pulley bears too hard on the end, or (6) the belt may be too tight.

Self-oiling boxes are very desirable where they can be

used, as they have a supply of oil which is carried up to the top of the shaft by a chain or ring. It is necessary to replenish the supply of oil only at rather long intervals.

64. Electrical transmission.*—Power may be transmitted by converting mechanical energy into electrical energy by the dynamo, and after transmission to a distance be converted into mechanical energy again by the electric motor. This form of transmission has many advantages where the electric current is obtained from a large central station, and no doubt will be an important form of transmission to the farmer of the future, as electric systems are spread over the country for various purposes.

*See Chapter XXII., Part II.

CHAPTER III

MATERIALS AND THE STRENGTH OF MATERIALS

A knowledge of the materials used in the construction of farm machinery and the strength of these materials will be helpful in the study of farm machinery.

65. Wood.—At one time farm machinery was constructed almost entirely with wooden framework, but owing to the increase in the cost of timber and the reduction in the cost of iron and steel, it has been superseded largely by the latter. Progress in the art of working iron and steel, making it more desirable for many purposes, has also been a factor in bringing about the substitution of iron and steel for wood. The woods chiefly used in the construction of farm machinery are hickory, oak, ash, maple. beech, poplar, and pine. It is not possible to discuss to any length the properties of these woods. The wood used in the construction of machinery must be of the very best, for there is no use to which wood may be put where the service is more exacting or severe. Wood used in farm machinery must be heartwood and cut from matured trees. It should be dry and well seasoned, and protected by paint or some other protective coating. Moisture causes wood to swell, and for this reason it is difficult to keep joints made of iron and wood tight, for the iron will not shrink with the wood.

Excessive moisture in wood greatly reduces its strength, and wood subjected to alternate dryings and wettings is sure to check and crack. Wood is especially well adapted to parts subject to shocks and vibrations, as the pitman of a mower. Iron, and especially steel, when subjected to shocks tends to become crystallized. This reduces its strength very much.

66. Cast iron is used for the larger castings and most of the gears used in farm machines. At one time cast iron was used to a larger extent than at the present time, as it is being superseded by stronger but more expensive materials. Cast iron is of a crystalline structure and cannot be forged or have its shape changed in any other way than by the cutting away of certain portions with machine tools. Cast iron has a high carbon content, but the carbon is held much as a mechanical mixture rather than in a chemical combination.

67. Gray iron is the name applied to the softer and tougher grade of cast iron, which is easily worked by tools; and **white iron** to a very hard and brittle grade. White iron is used for pieces where there are no changes to be made after casting.

68. Chilled iron.—When it is desired to have a very hard surface to a casting, as the face of a plow, the inside of a wheel box, or other surfaces subjected to great wear, the iron is chilled when cast by having the molten iron come in contact with a portion of the mold made up of heavy iron, which rapidly absorbs the heat. Chilled iron is exceedingly hard.

69. Malleable iron is cast iron which has been annealed and perhaps deprived of some of its carbon, changing it from a hard, brittle material to a soft, tough, and somewhat ductile metal. The process of decarbonation usually consists in packing castings with some decarbonizing agent, as oxide of iron, and baking in a furnace at a high temperature for some time. Malleable iron is much more expensive and more reliable than common cast iron.

70. Cast steel.—The term cast steel, as usually applied to the material used in the construction of gears, etc., is cast iron which has been deprived of some of its carbon before being cast.

71. Mild and Bessemer steel.—It is from this material that agricultural machinery is largely constructed. The hardness and stiffness of Bessemer steel varies and depends largely upon the carbon content. Steel with a high per cent of carbon (0.17 per cent) is spoken of as a high-carbon steel, and steel with a low per cent (0.09 per cent) low-carbon steel. Bessemer steel is difficult to weld.

72. Wrought iron.—Wrought iron is nearly pure iron, and is not as strong nor as stiff as mild steel, but can be welded with greater ease.

73. Tool steel is a high-carbon steel made by carbonizing wrought iron, and owing to the carbon content may be hardened by heating and suddenly cooling. Tool steel is used for all places where cutting edges are needed.

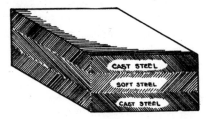

FIG. 27—DRAWING ILLUSTRATING THE CONSTRUCTION OF SOFT-CENTER STEEL

74. Soft-center steel, used in tillage machinery, is made up of a layer of soft steel with a layer of high-carbon steel on each side. The high-carbon steel may be made glass hard, yet the soft center will support the surface and prevent breakage. In making soft-center steel, a slab of high-carbon steel is welded to each side of a soft steel slab and the whole rolled into plates (Fig. 27). A soft-center steel may be made by carbonizing a plate of mild steel by a process much the reverse of malleable making.

STRENGTH OF MATERIALS

All materials used in construction resist a stress or a force tending to change their form. Stresses act in three ways: (1) tension, tending to stretch; (2) compression, tending to shorten; and (3) shear, tending to slide one portion over another.

75. Tension.—Material subjected to a stress tending to stretch it, as a rope supporting a weight, is said to be under tension, and the stress to the square inch of the cross section required to break it is its tensile strength.

76. Compression. — Material is under compression where the stress tends to crush it. The stress to the square inch required to crush a material is its compressive strength.

77. Shear.—The shearing strength of a material is the resistance to the square inch of cross section required to slide one portion of the material over the other.

78. Transverse strength of materials.—When a beam is supported rigidly at one end and loaded at the other, as in Fig. 28, the material of the under side of the beam is under a compressive stress, and that of the upper part is subjected to a tensile stress. The property of materials to resist such stresses is termed their transverse strength.

79. Maximum bending moment (B.M.) is a measure of the stress tending to produce rupture in a beam, and for a cantilever beam (i. e., one supported rigidly at one end, Fig. 28) is equal to the load times the length of the beam $(W \times L)$. The maximum bending moment depends upon the way a beam is loaded and supported; thus with a simple beam loaded at the center and supported at both ends the bending moment is one-half the weight one-half times the length.

The maximum bending moment for the cantilever beam of Fig. 28 is at the point where it is supported. If the beam be of a uniform cross section, it will rupture at this point before it will at any other. The bending

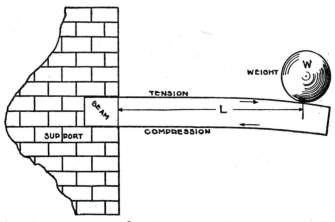

FIG. 28—A CANTILEVER BEAM

moment in the beam at hand grows less as the distance from the weight becomes less. As the bending moment becomes less, less material is needed to resist it, and hence a beam may be designed of such a section as to be of equal strength at all points, or it is what is called a beam of uniform strength.

Much material may be saved by placing it where most needed. The location as well as the value of the maximum bending moment depends upon the way the beam is loaded.

80. Modulus of rupture (M.R.).—It is seldom that a material has a tensile strength equal to its strength to resist compression, so neither of these may be used for transverse stresses. The modulus of rupture is a measure of the transverse stresses necessary to produce rupture

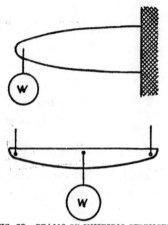

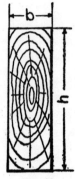

FIG. 29—BEAMS OF UNIFORM STRENGTH   FIG. 30

and is determined experimentally. It is usually a quantity lying between the compressive and tensile strengths of the material.

81. Section modulus (S.M.) is the quantity representing the ability of the beam to resist transverse stresses. It has been noticed by all that a plank will support a greater load on the edge than on the flat. For a rectangular cross section, Fig. 30, if h = depth in inches and b = breadth in inches, the section modulus is

$$\frac{b\,h^2}{6};$$

that is, the strength of a rectangular beam is proportional to its breadth and to the square of its depth.

*When a beam is loaded to its limit, bending moment =
section modulus × modulus of rupture.*

This is a general equation which applies to all beams.

82. Factor of safety.—In the design of machinery it is
customary to make the parts several times as strong as
would be needed to carry normal loads. The number
of times a piece is made stronger than necessary simply
to carry the load is called the factor of safety, and in
farm machine design it varies from 3 to 12.

For a more complete discussion of this subject see any
work on mechanics of materials.

AVERAGE STRENGTH OF MATERIAL PER SQUARE INCH

Material	Tensile Strength	Compressive Strength	Modulus of Rupture
Hickory......................		9,000	15,000
Oak		8,500	13,000
White pine..................		5,400	7,900
Yellow pine................		8,000	10,000
Cast iron...................	18,000	80,000	45,000
Steel........................	60,000	52,000	55,000
Wrought iron..............	50,000	48,000	48,000

Values for the strength of timber were obtained from
U. S. Forestry Circular No. 15. If the load or stress be
continued for a long time the ultimate strength of timber
will be only about one-half the above and for this reason
much lower values are often given in architects' hand-
books.

For a more complete table see any engineers' handbook.*

*"Architects' and Builders' Pocket-Book." By F. E. Kidder.
"Materials of Construction." By J. B. Johnson.

Problem: Find the safe load on an oak doubletree 4 feet long, 4 inches wide, 2 inches thick. Factor of safety = 6.

Let L = length in inches, W = load in pounds, b = thickness, d = width in inches.

Bending moment = $\frac{1}{4}WL = \frac{1}{4}W48 = 12W$.

Section modulus = $\frac{bd^2}{6} = \frac{2 \times (4)^2}{6} = 5.333$.

Modulus of rupture for oak = 13,000.

Bending moment = $\frac{\text{Sect. Mod.} \times \text{Mod. of Rupt.}}{\text{Factor of Safety}}$

$12W = \frac{5.333 \times 13,000}{6}$

$W = 963$ pounds. (Ans.)

CHAPTER IV

TILLAGE MACHINERY

83. Object of tillage.—Agricultural implements and machines used in preparing the soil for the seeding or growth of crops may be classed as tillage machinery. Tillage is the art which includes all of the operations and practices involved in fitting the soil for any crop, and the caring for it during its growth to maturity.

Tillage is practiced to secure the largest returns from the soil in the way of crops. Its objects have been enumerated in other works about as follows:

(1) To produce in a field a uniform texture to such a depth as will render the most plant food available.

(2) To add to the humus of the soil by covering beneath the surface to such a depth as not to hinder further cultivation, green crops and other vegetable matter.

(3) To destroy and prevent the growth of weeds, which would tend to rob the crops of food and moisture.

(4) To modify the condition of the soil in such a way as to regulate the amount of moisture retained and the temperature of the soil.

(5) To provide such a condition of the soil as to prevent excessive action of the rains by washing and the wind by drifting.

At the present time practically all of the various operations of tillage are carried on by aid of machinery, and for this reason tillage machinery is of greatest importance in modern farming operations. Modern tillage machinery has enabled the various objects as set forth to be realized, thus not only increasing the yield an acre, but at the same time permitting a larger area to be tilled.

THE PLOW

84. The development of the plow.—The basic tillage operation is that of plowing, and for this reason the plow will be considered first. Some of the oldest races have left sculptural records on their monuments describing their plows. From the time of these early records civilization and the plow have developed in an equal proportion. The first plow was simply a form of hoe made from a crooked stick of the proper shape to penetrate and loosen the soil as it was drawn along. The power to draw the plow was furnished by man, but later, as animals were trained for draft and burden, animal power was substituted and the plow was enlarged.

The records of the ancient Egyptians illustrate such a plow. At an early time the point of the plow was shod with iron, for it is recorded that about 1,100 years B.C. the Israelites, who were not skilled in the working of iron, "went down to the Philistines to sharpen every man his share and his coulter." In the "Georgics," Virgil describes a Roman plow as being made of two pieces of wood meeting at an acute angle and plated with iron.

During the middle ages there was but little improvement over the crude Roman plow as described by Virgil. The first people to improve the Roman model were the Dutch, who found that a more perfect plow was needed to do satisfactory work in their soil. The early Dutch plow seems to have most of the fundamental ideas of the modern plow in that it was made with a curved moldboard, and was provided with a beam and two handles. The Dutch plow was imported into Yorkshire, England, as early as 1730, and served as a model for the early English plows. P. P. Howard was one whose name may be mentioned among those instrumental in the development of the early English plow. Howard established a factory, which remains to this day.

James Small, of Scotland, was another who did much toward the improvement of the plow. Small's plow was designed to turn the furrows smoothly and to operate with little draft.

Robert Ransome, of Ipswich, England, in 1785 constructed a plow with the share of cast iron. In 1803 Ransome succeeded in chilling his plows, making them very hard and durable. The plows of Howard and Ransome were provided with a bridle or clevis for regulating the width and depth of the furrow. These plows were exhibited and won prizes at the London and the Paris expositions of 1851 and 1855.

85. American development.—Before the Revolutionary War the plows used in America were much like the English and Scotch plows of that period. Conditions were not favorable to the development of new machinery or tools. The plow used during the later colonial period was made by the village carpenter and ironed by the village smith with strips of iron. The beam, standard, handles, and moldboard were made of wood, and only the cutting edge and strips for the moldboard were made of iron.

Among those in America who first gave thought to the improvement of the plow was Thomas Jefferson. While representing the United States in France he wrote: "Oxen plow here with collars and harness. The awkward figure of the moldboard leads one to consider what should be its form." Later he specified the shape of the plow by stating: "The offices of the moldboard are to receive the sod after the share has cut it, to raise it gradually, and to reverse it. The fore end of it should be as wide as the furrow, and of a length suited to the construction of the plow."

Daniel Webster is another prominent American who, history relates, was interested in the development of the plow. He designed a very large and cumbersome plow for use upon his

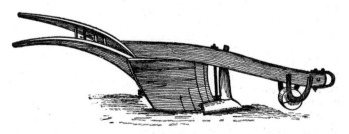

FIG. 31—WEBSTER'S PLOW

farm at Marshfield, Massachusetts. It was over 12 feet long, turned a furrow 18 inches wide and 12 inches or more deep, and required several men and yoke of oxen to operate it.

Charles Newbold, of Burlington, New Jersey, secured the first letters patent on a plow in 1797. Newbold's plow differed from others in that it was made almost entirely of iron. It is stated

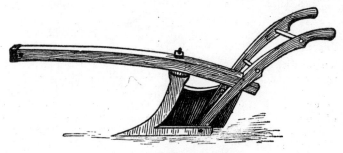

FIG. 32—THE NEWBOLD PLOW

that the farmers of the time rejected the plow upon the theory that so much iron drawn through the soil poisoned it, and not only retarded the growth of plants, but stimulated the growth of weeds.

Jethro Wood gave the American plow its proper shape. The moldboard was given such a curvature as to turn the furrow evenly and to distribute the wear well. Although Wood's plow was a model for others which followed, he was unrewarded for his work, and finally died in want. William H. Seward, former Secretary of State, said of him: "No man has benefited his country pecuniarily more than Jethro Wood, and no man has been as inadequately rewarded."

86. The steel plow.—As farming moved farther west the early settlers found a new problem in the tough sods of the prairie States. A special plow with a very long, sloping moldboard was found to be necessary in order to reduce friction and to turn the sod over smoothly. Owing to the firmness of the sod, it was found that curved rods might be substituted for the mold-board. Later when the sod became reduced it was found that the wooden and cast-iron plows used in the eastern portion of the country would not scour well. This difficulty led to the use of steel in the making of plows. Steel, having the property of taking an excellent polish, permitted the sticky soils to pass over a moldboard made of it where the other materials failed.

In about 1833 John Lane made a plow from steel cut from an old saw. Three strips of steel were used for the moldboard and one for the share, all of which were fastened to a "shin" or frame of iron. John Lane secured in 1863 a patent on soft-center steel, which is used almost universally at the present time in the making of tillage tools. It was found that plates made of steel were brittle and warped badly during tempering. Welding a plate of soft iron to a plate of steel was tried, and, although the iron supported the steel well when hardened, it warped very badly. The soft-center steel, which was formed by welding a heavy bar of iron between two bars of steel and rolling all down into plates, permitted the steel to be hardened without warping. It is very strong on account of the iron center, which will not become brittle.

In 1837 John Deere, at Grand Detour, Illinois, built a steel plow from an old saw which was much similar to Lane's first plow. In 1847 Deere moved to Moline, Illinois, and established a factory which still bears his name. William Parlin established a factory about the same time at Canton, which is also one of the largest in the country.

FIG. 33—THE MODERN STEEL WALKING PLOW WITH STEEL BEAM AND MOLD-BOARD FOR STUBBLE OR OLD GROUND

87. The sulky or wheel plow.—The development of the sulky or wheel plow has taken place only recently. F. S. Davenport invented the first successful sulky plow, i. e., one permitting the operator to ride, February 9, 1864. A rolling coulter and a three-horse evener were added to this by Robert Newton, of Jerseyville, Illinois. But E. Goldswait had patented a fore carriage in 1851 and M. Furley a sulky plow with one base December 9, 1856. Much credit for the early development of the sulky plow is due to Gilpin Moore, receiving a patent January 19, 1875, and W. L. Cassady, to whom a patent was granted May 2, 1876. Cassady first used a wheel for a landside. Too much space would be required to mention the many inventions and improvements which have been added to the sulky plow.

FIG. 34—AN UNDER VIEW OF THE MODERN STEEL PLOW, SHOWING ITS CONSTRUCTION

88. The modern steel walking plow.—Fig. 34 shows the modern steel walking plow suitable for the prairie soils. The parts are numbered in the illustration as follows:

1. Cutting edge or share. The **point** is the part of the share which penetrates the ground, and the **heel** or **wing** is the outside corner. A share welded to the landside is a **bar** share, while one that is independent is a **slip** share.

2. Moldboard: The part by which the furrow is turned. The **shin** is the lower forward corner.

3. Landside: The part receiving the side pressure produced when the furrow is turned. A plate of steel covers

the landside bar, furnishing the wearing surface. When used for old ground, the plow is usually constructed with the bar welded to the frog, forming the foundation to which the other parts are attached. Landsides may be classed as high, medium, and low.

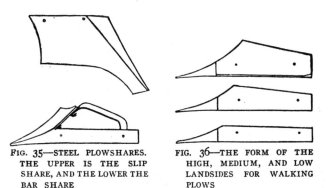

FIG. 35—STEEL PLOWSHARES. THE UPPER IS THE SLIP SHARE, AND THE LOWER THE BAR SHARE

FIG. 36—THE FORM OF THE HIGH, MEDIUM, AND LOW LANDSIDES FOR WALKING PLOWS

4. Frog: The foundation to which are attached the share, moldboard, and landside.

5. Brace.

6. Beam: May be of wood or steel. The beam in a wooden-beam plow is joined to the plow by a beam standard.

7. Clevis, or hitch for the adjustment of the plow.

8. Handles: The handles are joined to the beam by braces.

9. Coulter: Classified as rolling, fin, or knife coulters.

89. **Material.**—While in the cheaper plows the moldboard and share may be of Bessemer or a grade of cast steel, in the best plows these and also the landside are usually made of soft-center steel or chilled iron. The beam is usually of Bessemer steel, while the frog may be of forged steel, malleable iron, or cast iron.

90. Reënforcements.—A patch of steel is usually welded upon the shin, the point of the share, and the heel of the landside. These parts are also made interchangeable so new parts may be substituted when worn.

91. Size.--Walking plows are made to cut furrows

FIG. 37—ROLLING CASTER AND ROLLING STATIONARY COULTERS, FIN HANGING, KNEE, DOUBLE ENDER, AND KNIFE CUTTERS OR COULTERS

from 8 to 18 inches. A plow cutting a 14-inch furrow is considered a two-horse, and one cutting a 16- or an 18-inch furrow a three-horse plow.

92. The modern sulky plow.—The name sulky plow is used for all wheel plows, but applies more particularly to single plows, while the name **gang** is given to double or larger plows. Fig. 38 illustrates the typical sulky plow, and reference is made to its various parts by number:

1. The moldboard, share, frog or frame, and landside is called the plow **bottom**. Most sulky plows are made with interchangeable bottoms, so it is possible to use the same carriage for various classes of work by using suitable bottoms.

2 and 3 are the **rear** and the **front furrow wheels,** respectively. These wheels are set at an angle with the

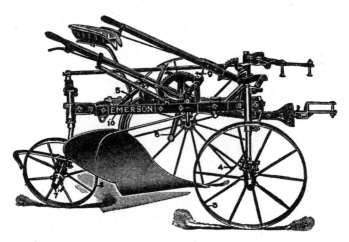

FIG. 38—THE MODERN FOOT-LIFT BEAM-HITCH SULKY PLOW WITH STEEL PLOW BOTTOM

vertical in order that they may carry to better advantage the side pressure of the plow due to turning the furrow slice.

4. The largest wheel traveling upon the unplowed land is spoken of as the **land wheel.**

5. The connections between the plow beam and the frame are called the **bails.**

6. A rod called the **weed hook** is provided to collect the tops of high vegetation.

7. Practically all wheel plows are now provided with **inclosed wheel boxes,** which exclude all dirt and carry a large supply of grease. The inclosed wheel box has a collar which excludes the dirt at the axle end of the wheel box, and has the other end entirely inclosed with a cap. The grease is usually stored in the cap, which is made detachable from the hub.

8. Wheel plows are now generally provided with a **foot lift,** by which the plow is lifted out and forced into the ground.

9. For plowing in stony ground, it is necessary to set the plow to float, so that in case a stone is struck the plow will be free to be thrown out of the ground without lifting the carriage, otherwise the plowman will be thrown from his seat and the plow damaged.

10. The various parts of the sulky plow are usually attached to the **frame,** and this is an important part in the construction of the plow. Not all sulky plows, however, are made with a frame.

93. Types of sulky plows.—Sulky plows differ much in construction. The two-wheel plow is not used extensively at the present time because it does not carry the side pressure of the plow well and does not turn a good square corner. One type of construction is that of a **frame** with wheels attached by means of brackets, making a carriage. To this carriage the plow proper is attached by bails. The **hitch** to frame plows may be to either the frame or to the plow beam. The former is known as a **frame hitch** and the latter as a **beam hitch.** There are good plows upon the market with a frame hitch, but the beam hitch plow seems to be preferred.

A cheaper type of plow than the frame plow is the

frameless, with the wheel brackets bolted directly to the plow beam. Such plows will often do very satisfactory work, but are not quite so handy. Frame plows are generally **high-lift** plows in that the plow may be lifted several inches above the plane of the carriage. A high-lift plow offers an advantage for cleaning and transporting from field to field.

With the cheaper plows there is no attempt to guide or steer the plow other than let it follow the team. Such plows may be classed as **tongueless.** A tongueless plow may, however, be provided with a hand lever either to shift the hitch or guide the front furrow wheel. Such a plow may be called a hand-guided plow, and the lever for guiding or adjusting is called the **landing** lever.

There is still another type of frameless plow wh ch is guided by the hitch. In the **hitch-guided** plow the front

FIG. 39—THE MODERN GANG PLOW

furrow wheel or the front and rear furrow wheels are steered by a connection to the plow clevis. A tongue may be used with this type of plow to keep the team straight and to hold the plow back from off the horses' heels while being transported.

The higher class sulky plows are guided with an adjustable tongue, the tongue being connected to the front and rear furrow wheels.

Sulky plows are usually fitted with a 14-, 16-, or 18-inch plow bottom, the 16-inch being the common size.

94. Gang plows.—Nearly every sulky plow upon the market has its mate among the gang plows, which, as stated before, do not differ greatly from it, only in that they have two or more plow bottoms instead of one. Gang plows usually have a hand lever to assist the foot lift in raising and lowering the plow. The common sizes of gang-plow bottoms are 12- and 14-inch.

FIG. 40—TYPES OF PLOW BOTTOMS. NO. 1 IS THE STUBBLE OR OLD GROUND BOTTOM. NO. 7 IS THE BREAKER BOTTOM FOR TOUGH NATIVE SODS. NOS. 2, 3, 4, 5, AND 6 ARE INTERMEDIATE TYPES FOR GENERAL PURPOSE PLOWS

95. Types of plows' bottoms.—The plow bottom, as stated before, is the plow proper, detached from the beam or standard. Owing to the varying conditions under which ground is to be plowed, a few general types, each with its own form of moldboard and share, have been established. These forms are illustrated in Fig. 40, and vary from No. 7, the **breaker,** with its long sloping share and moldboard, for natural sods, to No. 1, the stubble plow with short, abrupt moldboard for old ground. The

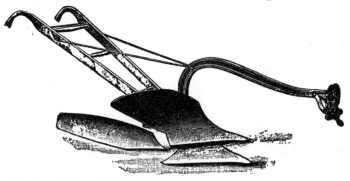

FIG. 41—A STEEL WALKING PLOW WITH INTERCHANGEABLE MOLDBOARDS, BY WHICH IT MAY BE MADE INTO A BREAKER OR STUBBLE PLOW

intermediate forms are given the name of turf and stubble, or **general purpose,** plows, being used for the sod of the cultivated grasses or for stubble ground. The breaker is suitable for the native sods of the Western prairies, as it turns the furrows very smoothly and covers the vegetation completely, that it may decay quickly. The abrupt curvature of the moldboard in the stubble bottom causes the furrow slice to be broken and crumbled in making the sharp turn, and thus has a more pulverizing action and is designed for old ground. The general purpose plow is designed for the lighter sods, such as those of the tame grasses.

Some manufacturers make plows with interchangeable moldboards, and sulky plows are usually built with interchangeable bottoms, so the plow or carriage may be used for a variety of soils.

96. The jointer.—The jointer is used in soils inclined to be soddy. It enables the plow to do cleaner work and cover all vegetation, throwing a ribbon-like strip of turf into the furrow. It will often render excellent service

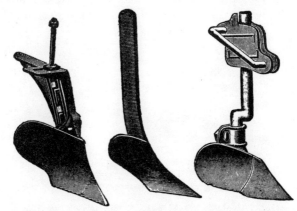

FIG. 42—TYPES OF JOINTERS. THE TWO AT THE LEFT ARE MADE OF STEEL; THE ONE AT THE RIGHT IS A CHILLED IRON JOINTER WITH AN ADJUSTABLE SHANK

where sod ground is to be plowed deep and left in shape for immediate pulverizing to fit it for crops. It will cut out a section of the sod, turning it into the bottom of the furrow, where it will be completely covered, and at the same time leave the upper edge of the furrow slice composed only of comparatively loose earth. By cutting out the corner of the furrow slice, the furrows will be completely inverted, leaving the surface smooth. If the furrow slice is perfectly rectangular, the furrows are inclined to pile or lap over each other.

97. The chilled plow.—In many places, especially in the eastern United States, many of the plows used are of chilled cast iron. A chilled plow with an interchangeable

FIG. 43—A MODERN CHILLED WALKING PLOW WITH JOINTER AND GAUGE WHEEL

point is shown in Fig. 43. Chilled plows are very hard, but will not scour in all soils. The share can only be ground to an edge when dull, or it may be replaced at a small cost.

98. The hillside plow.—In localities too sloping to throw the furrow uphill, hillside or reversible plows are

FIG. 44—A REVERSIBLE OR HILLSIDE PLOW WITH KNIFE COULTER

used. A plow which may be made a right- or left-hand plow by turning it under on a hinge to the standard is shown in Fig. 44. In irrigated districts where dead furrows interfere with the carrying of water upon the land, reversible plows are used. These are of many forms, but the type will not be further discussed.

99. The subsoil plow.—Where it is desirable to loosen the ground to a greater depth than can be done with a surface plow, the subsoil plow is used. It is used with

FIG. 45—A SUBSOIL PLOW FOR LOOSENING THE SOIL IN THE BOTTOM OF THE FURROW MADE BY THE COMMON PLOW

the regular plow, following in the furrow made by it. Opinions in regard to the value of this plow differ, but the subject will not be discussed here.

100. The disk plow.—The disk plow is the result of an effort on the part of inventors to reduce the draft due to the sliding friction upon the moldboard. Figs. 46 and 47 show the modern disk plow made for horse and engine power, respectively. A plow consisting of three disks cutting very narrow strips was about the first one patented, M. A. and I. N. Cravath, of Bloomington, Illinois, being its inventors. Under certain conditions, it is said, this plow did very satisfactory work, but the side pressure was not sufficiently provided for. M. F. Hancock succeeded in introducing the disk plow into localities where conditions were well adapted to its use, and became prominent as a promoter of the disk plow.

FIG. 46—A DISK GANG PLOW TO BE OPERATED BY HORSE POWER

The draft of the disk plow is often heavier in proportion to the amount of work done, and the plow itself is

FIG. 47—AN ENGINE DISK GANG PLOW TURNING 8-, 10-, OR 12-INCH FURROWS

more clumsy than the moldboard plow; so where the latter will do good work there is no advantage in using the former. In sticky soils, however, or in very hard ground, where it is impossible to use the moldboard plow, the disk will often be found to do good work, and in the latter case with much less draft. The moldboard plow is recommended by the manufacturers of both plows where it will do good work.

Disk plows have been made in the walking style within the past few years, but have proved rather unsatisfactory. A few of this style are suitable for hillside and irrigation plows, being made reversible.

101. The steam plow.—Where steam power is used for other purposes, or where farming is carried on extensively, steam may be used at a saving over horse power in plowing. This has been attempted for many years, but it has only recently become very successful, and even now the steam plow is used only on large farms and on level land. If the soil is not firm, the great weight causes the traction wheels of the engine to sink into the ground until the plow cannot be pulled.

The modern steam plow, direct connected, steered from the rear, and having a steam lift, is a very successful machine. Its advantages are its capacity and unlimited power for deep plowing. The cost of plowing with a steam plow varies with the cost of fuel and other conditions, but it should be from 75 cents to $1.50 an acre. Outfits capable of plowing and at the same time preparing the seed bed and seeding 40 to 50 acres in a day are now in use.

A type of steam plow which has been successful in Europe is operated by a system of cables. The plow is drawn back and forth across the field by means of the cable, the engine being placed at one end of the field.

The steam plow may, in some cases, in certain soils, be the means of producing an increase of yield of crops, by plowing to a greater depth than could be done by horse power.

102. The set of walking plows.—The original set of a plow, or the proper adjustment of its point, share, and beam, is given by the maker. Each time when the plow is sharpened the smith is depended upon to return this set to the plow.

103. Suction.—The suction of a plow is usually measured as the width of the opening between the landside and a straight edge laid upon it when the plow is bottom side up. It is usually about ⅛ inch, but may vary slightly

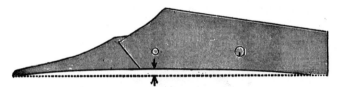

FIG. 48—THE SUCTION OF WALKING PLOWS SOMEWHAT EXAGGERATED

without detriment to the plow. It may also be described as the amount the point is turned down to secure penetration.

The point of the share is also turned slightly outward, which makes the line of the landside somewhat concave. The beam of a three-horse plow is in a line with the landside, but in a two-horse plow it is placed a little to the furrow side of the line of the landside, usually about 3 inches, in order that the hitch may be more directly behind the team. For ordinary plows the point of the beam stands 14 inches high, but it is higher for hard soils. Some **bearing** must be given at the heel of the share in **walking** plows, to carry the downward pressure of the furrow. One inch width of bearing surface for 12- and 14-inch plows and 1¼ inches for 16-inch plows is the average width of this bearing, more being needed for soft, mellow soils than for firm soils. This fact necessitates a change in the plow in changing from hard to mellow soils, as a share set for a hard soil will swing to one side or work poorly in the mellow soil. A handy device called a **heel plate** is sometimes used to vary the width of surface at the heel.

104. The set of sulky plows.—With the sulky plow, when the share lies on a flat surface, the distance from the heel of the landside to the surface is called the suction.

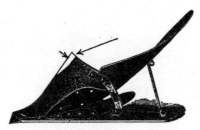

FIG. 49—THE BEARING SURFACE REQUIRED AT THE WING OF THE SHARE

In sulky and gang plows this is usually ½ inch. The entire downward pressure or suction should be carried upon the wheels or carriage, which, with their well lubricated bearings, will reduce the draft and require no bearing surface at the wing of the share. In order to reduce the friction by removing the pressure from the landside, the rear furrow wheel is set outside the line of the landside, usually about 1¼ inches.

105. Set of coulter.—The rolling coulter should be set to clear the shin of the plow by about ¼ inch, and should cut ½ inch or ¾ inch outside the shin. It is said that if the coulter is made to cut 1 inch or more outside the landside, thus increasing the load upon the plow, it can be made to scour when giving difficulty in this respect. When plowing among roots the plow is enabled to run over rather than underneath large roots by inclining the knife coulter backward with its point below the point of the plow; otherwise the knife coulter must be set with the lower point well ahead.

106. Scouring.—Some soils are of such a nature that a plow can be made to scour only with difficulty. This is true especially of soils in the Middle West. In other localities plows give little trouble in this respect. When the plow is at fault, poor scouring may be due (1) to poor temper. In this case the share and moldboard are not hard enough to take a good polish, and hence will not scour well. These parts should be so hard that they can barely be scratched with a file. (2) To poor grinding. Sometimes hollows have been ground into the moldboard, over which the furrow slice presses so lightly that not enough pressure is given to cause the spot to scour. This may readily be tested by carrying the tips of the fingers up the plow quickly, from the edge of the share in the direction the soil moves. (3) To a poor fitting, i. e., where the joint between the share and moldboard is not smooth. A remedy for this is procured by shimmering the share up or down with small pieces of pasteboard. (4) To the edge of the share not being level, making a low spot back of the edge. This is usually caused by a warped share. (5) To poor setting. The plow must be set as previously described.

107. Sharpening steel shares.—It is recommended by some manufacturers that until necessary only the extreme point of a share be heated to put into form, the edge being sharpened by grinding; but when necessary to heat and draw to an edge by hammering, they recommend the following procedure:

The point should be heated to a low cherry red. If the heat is too intense, the quality of the steel will be injured. Only as much should be heated at once as can be hammered. The body of the share must be kept cool and strong so the fitting edges may not be disturbed. After this, the entire cutting edge should be cold hammered. The share should then be set on a level platform, leaving 1/16 inch under the middle piece to give proper suction or pitch. The edge must touch all the way along, and the proper bearing must be given at the wing.

108. Hardening plowshares.—A hardened share will retain its cutting edge much longer than a soft share. It is highly advisable, after each time the edge is drawn out by heating and hammering, that the share be hardened. Some soils require hardened steel shares in order that they may retain their scouring qualities. Several reliable manufacturers give directions for sharpening and hardening shares made of soft-center steel about as follows: Sharpening: The whole point should be heated to a very low red heat, then the face of the share must be turned downward with the heel over the fire and the point about 2 inches higher than the heel. In this way the whole length of the share will be heated almost in one heat, as the fire will be drawn along from the heel toward the point. An uneven heat will warp and crack the share. When a moderate heat has been reached it must be removed, and it will be noticed if the share is sprung up along the edge. This must be set right, and the following methods may be used to harden:

First. The edge must be made hard and springy by cold hammering; then the share is to be heated as described to a low cherry red. It should be let into the water (holding it bottom side up) far enough to cool the edge, then taking it out, and the color should be watched as the heat returns to the edge. When a dark straw or mottled purple reaches the edge, the entire share may be cooled.

Second. If a supply of oil is at hand, the share may be tempered with less risk of breakage. When oil is used (linseed or lard oil will answer) the share is to be heated as before to a low cherry heat, then lowered into the oil till entirely cool. After this it must be held over the fire till the temper is sufficiently drawn, which will be indicated by the oil on the thin part of the share taking fire. It may finally be cooled by immersion in cold water.

109. Draft of plows.—The nature of soils, growth of roots, and amount of moisture present influence the draft of plows. The shape of the moldboard also affects the draft, the more abrupt curvature producing a more pulverizing action upon the furrow slice, and requiring more work.

Professor J. W. Sanborn, of Missouri, made tests to determine the reduction of draft due to the use of a coulter, the results of which are as here given. The tests were made with a plow similar to the sod or breaking plow, and in clover sod two years old, with about as much moisture present as would permit working the soil advantageously. The results were as follows:

	Size of Furrow	Total Draft	Draft per Sq. In.
Sod plow with wheel coulter...	5.575″ × 15.08″	296.25	3.524
" " without " ...	5.325″ × 14.5″	343.75	4.453
Difference............		47.50	.929

The coulter resulted in better work and diminished the draft 20.86 per cent. A later series of observations was made on clover sod, the plow being provided with a wheel coulter, the soil being drier than before. The following results were obtained:

	Size of Furrow	Total Draft	Draft per Sq. In.
Clover sod without coulter....	6.47″ × 11.61″	714.35	10.80
" " with "	6.413″ × 12.47″	664.82	8.616
Difference....................................		49.53	2.184

In these tests the coulter reduced the draft 25.34 per cent.

It is stated in the report of the trials of plows at Utica that the total draft of a plow is divided as follows: 35 per cent is used in overcoming the friction between the implement and the soil, 55 per cent in cutting the furrow slice, and 10 per cent in turning it. The accuracy of these tests has been doubted by some, but the tests seem·to have been conducted with care, and they show the necessity of keeping a sharp cutting edge. It is desirable that data of this kind be obtained by tests made with modern plows.

110. **Draft of sulky plows.**—It is often claimed that the draft of sulky plows is less than that of walking plows, owing to the friction of the sole and landside being transferred to the well-oiled bearings of the carriage. But records show that there is no gain unless the weight of the driver and the frame is deducted. But there is an evident advantage in riding plows, even if the draft is slightly greater on the team with the plowman riding rather than walking, and the plow being handled with equal facility. Though little information is at hand on the subject, what there is seems to indicate that there is only a slight difference in the draft of walking and riding plows, in proportion to the amount of work done.

111. **The selection of a walking plow.**—The best in

quality of material and workmanship is desirable when selecting a walking plow. It may be difficult to judge of the material, but the workmanship can be easily determined. Beginning with the frog, the plow should be well made and put together, and at this point a vast difference in plows may be detected. The work to be done should determine the kind of plow to be selected, and the type of mold-board must be suited to the soil to be turned. While steel-beamed plows are used to better advantage in plowing among trash, plows with wooden beams have an advantage in being lighter and less likely to be sprung. A wooden-beam plow, striking a rock or root, may have the beam broken, while with a steel-beam plow it may be distorted. A right-hand plow is one that turns the furrow to the right, and a left-hand plow is one turning the furrow to the left. The custom established in the locality where it is to be used should determine the one to select, as one has no advantage over the other.

112. The selection of a sulky plow.—As is the case with the walking plow, the quality of a sulky plow will be indicated largely by its construction and workmanship, although its selection requires more care than that of a walking plow. To be brief, a well-made plow and one easily operated as regards foot lifts and levers should be chosen. It should turn a square corner in either direction, and all parts subject to wear should either be adjustable or made of generous dimensions. This applies especially to bail boxes on bail plows.

113. Adjusting the walking plow.—A few points regarding the operation of plows should be mentioned. A walking plow, if working properly, should need very little attention from the plowman, only requiring him to steady it with the handles. If it requires a steady pull to either side, either the hitch or the clevis should be adjusted or the amount of bearing given at the heel or wing is too great or too small. It should be seen that the point is well turned down and never allowed to become rounding. If it becomes much worn, new metal must be added. It is desirable to maintain the original amount of suction and the distance from point of share to point of beam; in fact, the entire form of the plow should be maintained as nearly as possible in its original condition, providing it worked satisfactorily when new.

As given in former data, a large proportion of the draft is due to the cutting of the furrow. This shows the importance of keeping the cutting edge sharp. It has also been stated that if after being sharpened the share is hardened, the cutting edge will be retained longer.

114. **Adjusting the sulky plow.**—The land wheel of a three-wheel sulky or gang plow should travel directly to the front, but often, owing to bad adjustment, it is required to slip occasionally, because it is traveling at an angle with the direction of the plow's motion. The rear furrow wheel is usually given a small "lead" from the land, i. e., it is turned out a little from the unplowed land. This wheel should also be set an inch or so outside of the line of the landside, in order to remove the friction from this part as much as possible. The front furrow wheel is given "lead" from the land with the single plow, and toward the land when the team is hitched abreast on gangs. This difference in the latter case is because the line of draft is outside the line of work, and the plow is made to travel directly to the front by the front furrow wheel being turned in.

In any wheel plow the load should be carried as much as possible on the wheels in order to reduce the draft. There should be a reduction in draft when the entire load, due to lifting and turning the furrow slice, is carried upon the carriage wheels with the well-lubricated bearings, rather than upon the sole and landside of the plow, where all is sliding friction.

Care should be taken in hitching that the horses are not too much crowded or spread too much, as in either case good work cannot be done. When spread too much the team cannot travel directly to the front so well, and the line of draft is too far out to do good work. When crowded, the horses are working at a disadvantage, and the heat in warm weather will affect them more. When not in use, the polished surface of a plow should be protected from rust by a coat of heavy grease or "axle grease," and, like all other implements, it should be protected from the weather.

CHAPTER V

TILLAGE MACHINERY (Continued)

115. The smoothing harrow. — After plowing the ground, it is necessary to pulverize the soil very finely and to smooth it. The harrow is the implement used for this purpose, and it may be used also to cover seeds, to form a dust mulch for retaining moisture, and to kill weeds when they are beginning to grow.

116. Development.—Formerly the branch of a tree of a size to suit the power, whether man or animal, was used as a harrow. The limb chosen had small branches extending usually all to one side or the other, so as to lie flat when in use. Even until quite recently the brush harrow has been in use for covering seeds. An early type of harrow consisted of a forked limb with spikes in each arm, to which a cross arm was added later. This form was known as the "A" harrow. Until late in the sixteenth century a type of harrow devised by the Romans was the standard. This harrow was square or oblong, having cross bars with many teeth in them.

117. Classification.—Harrows may be classified as follows:

```
1. Smoothing harrows.
      Kinds of teeth........Straight fixed tooth;
                            Square-and-round tooth;
                            Cultivator tooth.
      Kinds of frame.......Wood frame;
                            Pipe frame;
                            Channel or U bar frame.
      Adjustment of teeth..Fixed tooth;
                            Adjustable tooth;
                            Lever harrows.
```

2. Spring-tooth harrows.
3. Curved knife-tooth harrows or pulverizers.
4. Disk harrows: Full disk; cutaway; spading; orchard.

It will not be possible to illustrate all these forms of harrows. The common **smoothing harrow** is not shown,

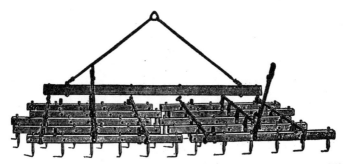

FIG. 50—A WOOD-BAR LEVER SMOOTHING HARROW. A CHEAPER HARROW
IS MADE WITH FIXED TEETH AND A WOODEN FRAME

but a lever harrow with wooden bars is shown in Fig. 50. Wooden-frame harrows can be used to better advantage in trashy ground when they are provided with a tooth fastener so arranged that the teeth will slope backward

FIG. 51—A CURVED KNIFE-TOOTH HARROW OR PULVERIZER

when drawn from one end. Such teeth may be spoken of as **adjustable.** A curved knife-tooth harrow, sometimes spoken of as a **pulverizer,** is illustrated in Fig. 51. This

FIG. 52—A RIDING WEEDER

crushes clods and brings the soil into uniform structure very satisfactorily. The weeder has rather long teeth and is an excellent implement for destroying small weeds, and also to form a dust mulch and a fine tilth. The culti-

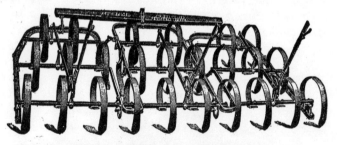

FIG. 53—A SPRING-TOOTH LEVER HARROW

vator tooth has the point flattened, and is curved so as to penetrate the ground more readily. Often it is aided in passing over obstacles by being held in place with a **spring.**

118. The spring-tooth harrow.—This harrow is illustrated in Fig. 53. When the teeth are caught on any obstacle they spring back and are released, this fact making it a very useful implement for stony ground. It is also an excellent pulverizer.

119. The selection of a tooth harrow.—It is a difficult matter to give explicit directions for selecting a harrow. The work to be done is the first thing to be considered, as a smoothing harrow, for instance, performs a very different office from a pulverizer or a weeder. Next the workmanship used in its manufacture and construction should be well examined. At all points where there will be much wear it should be well reënforced, and should have the general appearance of being a well-made tool. The connection between the sections of the evener especially should be properly reënforced, as the work of a single season has been known to wear out these connections. The tooth fastener is another important part in a tooth harrow which demands the attention of the purchaser. The tooth should have a head so that it will not drop out and be lost in case the fastener should become loosened. The square tooth is desirable, though spike teeth are made either from round or square stock. The regular sizes are ½ inch and ⅝ inch, the ⅝ inch size being suitable for heavier work. The number of teeth to the foot of the harrow may vary from five to eight, and this number as well as their size should correspond to the kind of work and conditions under which the harrow is to be used. Originally wooden harrow frames were the only kind used, but now they are generally made of steel pipe, angle and channel bars. The later styles of harrow are much more durable, and, the same amount of material being used, there is little choice between the styles of steel harrows. **Lever** harrows have an advantage in that the angle of the tooth may be adjusted, making the implement capable of performing a variety of work. Some levers are more easily operated than others. This lever adjustment facilitates transportation. Some harrows are so constructed that the sections may fold upon each other for easy transportation. Harrows in which the ends of the tooth bars are protected are suited for orchard work, as the bars will not catch and bark the trees.

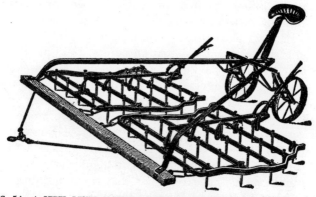

FIG. 54—A STEEL LEVER HARROW WITH A RIDING ATTACHMENT OR HAR-
ROW CART. THIS HARROW HAS THE TOOTH BARS MADE OF
STEEL CHANNEL BARS WITH PROTECTED ENDS

120. The harrow cart.—In order that the operator may ride, this device is sometimes attached behind the harrow. The attachment is made to the eveners by angle bars, and the wheels are made to caster so that in turning it will follow the harrow. It is very laborious to walk behind the harrow on plowed ground, and the harrow cart removes this difficulty; at the same time the rider has easy control of the team and is above the dust. The extra draft should be very little, but the wheels should have wide tires to prevent them from cutting into the soft ground.

121. The disk harrow.—This tool is perhaps the best adapted for pulverizing and loosening the ground of any yet devised. On account of its rolling action, it can be used for a great variety of conditions. It does excellent service in reducing plowed ground which is inclined to be soddy, and may even be used to prepare hard and dry soils for plowing. It may also be used to advantage in destroying weeds after they have grown beyond the control of the smoothing harrow. In fact, the disk harrow should be in use on every farm.

FIG. 55—A TWO-LEVER DISK HARROW. SCRAPERS OPERATED BY THE FEET

122. The full-bladed disk harrow.—This class of harrow may be used to good advantage as a pulverizer, and the blades are easily sharpened when dull, either by grinding or turning to an edge. The diameter of the disks may vary from 12 to 20 inches. For general purposes, the medium-sized, or 14- or 16-inch, disk is the size best adapted, although the larger sizes may have slightly less draft. The penetration of the disk blades into the ground is determined by (1) the line of draft, (2) the angle of gangs, (3) the curvature of the disk blades, (4) the weight of the harrow, and (5) the sharpness of the blades.

123. The cutaway or cut-out disk harrow.—As may be judged from the name, portions of the periphery of the blade of this harrow are notched out, allowing the remaining portions to penetrate the ground to greater

FIG. 56—A SINGLE-LEVER CUTAWAY DISK HARROW

depth. The entire surface, however, is not so thoroughly pulverized as with the full-bladed disk. It has a disadvantage of being hard to sharpen. The cutaway harrow seems to be especially adapted to work among stones and may be used to cultivate hay land.

124. Spading harrow.—This type of harrow has blades curving at the ends, forming a sort of sprocket wheel, with the cutting edges out. It works much like a cutaway. To sharpen it the blades must be separated and

drawn out much as a plow is sharpened. A special form
of spading harrow with sharp spikes is used in cultivating

FIG. 57—A SPADING HARROW

alfalfa, and is given the name of "alfalfa harrow." The
orchard disk differs from the common disk only in that
it has an extension frame, so that it may be used to

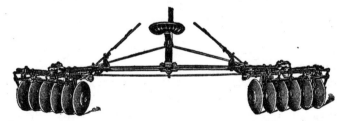

FIG. 58—AN ORCHARD DISK HARROW WITH WIDE FRAME TO WORK UNDER
TREES. THE GANGS MAY BE SET TO THROW IN OR OUT

cultivate rows of small plants as well as to reach under
trees and cultivate the soil under the branches. The disk

gangs often may be set to throw in or out from the center, to suit the nature of the work.

Usually the first parts of the disk harrow to wear out are the bearings. There are many styles of ball and chilled iron bearings in the market now, but those of hard wood seem to be as satisfactory as any, since they may be easily replaced. The construction of the bearings should be such as to exclude all dirt. A reliable means of oiling should be provided, and it is well to have an oil pipe to the bearings which extends above the weight pans or frame.

The scrapers or cleaners to keep the disks clean are another important feature of the disk harrow. These may be made stationary or so arranged as to be operated by the feet of the driver or otherwise when needed. They are not needed when working in dry soil, and when stationary they cause undue friction. A scraper that is made to oscillate by horse power over the face of the disk blades, and clean them automatically once in six revolutions, is sometimes used. When not needed it may be thrown out of gear.

Disk harrows with bumpers to carry the end thrust of the sections are usually made with one lever in order that the gangs or sections may be adjusted and the bumpers kept squarely together. A scheme to surmount this difficulty is to adjust the outer end of the gangs only. A two-lever disk harrow offers several advantages by adjusting the gangs at different angles for side hill work and for double disking by lapping one-half each time. The soil when disked once is not as firm as the undisked ground, and if lapping one-half, it may be necessary to set the gangs at different angles in order to cause the harrow to follow the team well.

It is advisable to have good clearance between standards and the disks and between the weight boxes and the disks. Good clearance will prevent clogging in wet and trashy ground. In order to secure flexibility of the gangs it is almost essential to have spring pressure to keep the inside ends of the gangs down. There is a natural tendency for the gangs to raise at the center. If three horses are to be used, it is advisable to have a stub tongue and an offset pole. Patent three-horse eveners to remove side draft with the pole set in the center are not to be advised. A liberal amount of material must be used in the construction as well as good workmanship—for instance, a heavy gang bolt with a lock nut. A square gang bolt is considered better than a round one.

125. Tongueless disk harrows are now made with a truck under a stub-tongue. These harrows, no doubt, make the work lighter for the team, but sacrifice a certain amount of control in handling the harrow. This feature is of more importance under certain conditions than others. A tongue truck is also used and is a very satisfactory addition to the harrow.

126. Plow-cut disk harrows.—Harrows have been constructed for several years with disks which have a raised or bulging center, the idea being that the dirt in being forced up over the raised center is turned over much like it would be from the moldboard of a plow. It is claimed by the manufacturer that this shape enables the harrow to cover the small trash better, that it leaves the ground leveler, and the harrow has better penetration on account of the shape of the disk blades. All these claims are denied by other manufacturers.

THE ROLLER AND PLANKER

127. The land roller is a very efficient tool for working up a fine tilth and for making the ground smooth and firm. The first rollers were constructed out of suitable logs and were drawn by yokes engaging pins in the ends of the rollers. It was soon found that if a log of any width was used, it would not work well on uneven ground, and it was clumsy to turn. Rollers made in two or three sections were then introduced, which were found in a great measure to overcome these difficulties. If the soil moisture is to be conserved, the roller should be followed by a smoothing harrow,

FIG. 59—A SMOOTH IRON ROLLER

as the former smooths and packs the ground, permitting the escape of the capillary water into the air. The harrow will roughen the surface, thereby decreasing the wind velocity, and will also put a dust mulch over the surface. The ground will be in much better condition for a mower or other machine after the roller has passed over it.

Certain advantages over the plain smooth rollers are claimed for the corrugated or tubular rollers, several styles of which have been invented. They are said to

FIG. 60—A TUBULAR ROLLER

crush the clods better, and they do not leave a smooth surface. Figs. 60 and 61 illustrate two rollers of this

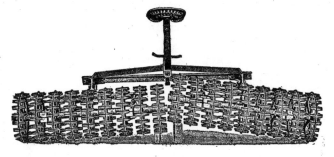

FIG. 61—A FLEXIBLE ROLLER AND CLOD CRUSHER OF SPECIAL DESIGN

type. H. W. Campbell invented a tool of this nature called the subsurface packer, for packing the ground beneath the surface. This tool (illustrated in Fig. 62) consists of a series of wheels with wedge-shaped tread.

Campbell advocates a method of surface cultivation to conserve the moisture in semi-arid regions. The intertillage of wheat and other small grains is included in this system. An authority states that rollers should be at least 2 feet in diameter, and should not weigh more than

FIG. 62—THE SUBSURFACE PACKER

100 pounds to the foot of width. If intelligently used, the roller is no doubt a valuable implement to the average farm.

128. The common planker, although a home-made tool,

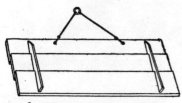

FIG. 63—THE COMMON PLANKER, A SERVICEABLE TOOL USUALLY MADE ON THE FARM

is a very valuable implement for crushing clods and smoothing the surface. It is not inclined to push surface clods into the soil like the roller, but will catch them and pulverize t h e m. The planker does not adapt itself well to any unevenness of the surface and does not pack the soil like the roller.

THE CULTIVATOR

129. Development.—The modern cultivator, which is a very efficient aid to the cultivation of growing plants, has developed under the addition of animal power from a kind of crude hoe used in the early days. The original single shovel was changed for the double shovel, this in turn was supplanted by the straddle-row cultivator, and even the latter was increased in size until in some cases the modern cultivator will take two rows at a time. A horse hoe and drill was invented by Jethro Tull early in the eighteenth century, but this was never a popular machine. Until 1860 country blacksmiths generally made the double shovels used by farmers. A patent was granted to George Esterly, April 22, 1856, on a straddle-row cultivator for two horses, and his was the first of the line of implements in the manufacture of which millions are now invested.

130. Classification of cultivators.

> Single- and double-shovel cultivators.
> One-horse cultivators.
> > Five- and nine-tooth cultivators.
> Straddle-row cultivators.
> > Walking—
> > > Tongue,
> > > Tongueless.
> > Riding.
> > Combined.
> > Single-row.
> > Double-row.
> Surface cultivators.

131. Single- and double-shovel cultivators, although used very extensively at one time, have their use confined almost entirely to garden and cotton culture.

132. The one-horse cultivator is used largely in gardening and for cultivating corn too high to be cultivated with the straddle-row cultivator. It may be provided with almost any number of teeth from 5 to 14. The teeth may vary from the harrow tooth designed for producing a very fine tilth, to the wide reversible shovels used on

the five-tooth cultivators. Also a spring tooth may be
used similar to those used on the spring-tooth harrow.

**133. Features of cultivators, with suggestions in regard
to selection.**—The **gangs** (sometimes called rigs) are the
beams, shanks, and shovels. Usually several styles of
gangs may be fitted to each cultivator. The shovels may
vary in number from four to eight for a pair of gangs.
The larger number is to be preferred•for producing the

FIG. 64.—FIVE- AND ELEVEN-TOOTH ONE-HORSE CULTIVATORS. EACH HAS
A LEVER FOR VARYING THE WIDTH, AND ALSO GAUGE WHEELS.
ONE HAS A SMOOTHING ATTACHMENT

proper tilth of the ground, but are very troublesome in
being easily clogged with trash. The six-shovel gangs
are very popular for corn culture. The eight-shovel
gangs may have each set of four shovels arranged either
obiquely or in what is called a zigzag. Best cultivator
shovels are made of soft-center steel. They are made of
almost any width, and may be straight or twisted. The
twisted shovel has a plow shape designed to throw the
dirt to one side or the other, while the straight shovel
must be adjusted on its shank to do this. The **beam** may
be made of wood, steel channel, flat bar, or pipe. The
wood beam is somewhat lighter, but not so strong or

durable. The **shanks** may be constructed of the same material as the beam and are provided either with a break-pin device or knuckle joint to prevent breakage when an obstruction is struck.

Flat springs may be used for the shanks, and when so used the term **spring tooth** is applied. Gopher shovels are arranged to take the place of a special surface cultivator. Such an arrangement is not generally satisfactory. A device is sometimes added to keep the shovels facing directly to the front. Such a gang is spoken of as having a **parallel beam.**

Seats are of two styles: the **hammock** and the **straddle.** The hammock seat is supported by the frame at each side and offers a good opportunity to guide the gangs with the feet. The straddle seat is more rigid, hence is well adapted to the treadle- or lever-guided cultivators.

The **pivotal tongue** is a device enabling the operator to vary the angle with which the tongue is attached to the cultivator frame. It may be used as a steering device, or to set the tongue at such an angle that the cultivator will not follow directly behind the team. It is very useful in side hill work where the cultivator tends to crowd down the hill. It may also be used in turning in a limited space.

The **expanding axle** permits the width of track to be varied, necessary on account of various widths of rows. It is accomplished by a divided steel axle or by the use of collars upon the axles. The divided axle permits of the use of the inclosed wheel box. It is an advantage to have the half axles reversible in that when the axle end becomes worn the opposite end may be substituted.

Spacing.—Some provision should be made to widen or narrow the spacing of the gangs or rigs. On single-row cultivators this is accomplished by slipping the couplings in and out upon the front arch. The spacing in two-row machines should be accomplished by a lever which permits the change to be made while in operation.

Suspension.—The gangs should be so suspended as to swing freely in a horizontal plane. If the point of suspension is too far back and the suspending arm or chain too short, the shovels will be lifted out of the ground as

FIG. 65—A TONGUELESS FOUR-SHOVEL CULTIVATOR WITH WOODEN GANGS. THE SHOVELS ARE NOT IN PLACE

the gang is carried to either side. The farther ahead the gang is suspended and the longer the suspending arm, the more nearly the gang will swing in a plane. Considerable difference is experienced in the ease with which a long gang is guided compared with a short gang. This is due to the fact that as a short gang is swung to one side more work is done, as the shovels must be carried ahead; while with a long gang the shovels are not carried ahead to such an extent.

Coupling.—The double hinge joint by which the cultivator gang is attached to the frame is called the coupling. Due provision should be found in the coupling for taking up wear. It is impossible to guide properly a gang with much lost motion in the coupling.

Raise of rigs.—Springs should be provided to aid the operator in lifting the heavy rigs. Also these springs are often used to aid in forcing the shovels into the ground.

Levers.—In riding cultivators the lifting levers should

FIG. 66—A RIDING BALANCE-FRAME FOUR-SHOVEL CULTIVATOR WITH HAM-
MOCK SEAT AND STEEL GANGS

be so placed as to be easily handled from the seat. In two-row machines it is very essential to be able to work each gang independently in raising and lowering. In this way one gang may be freed from trash without molesting the others.

Balance frame is a name applied to cultivators so constructed that the position of the wheels may be so adjusted, either by a lever for the purpose or by the movement of the gangs, as to balance the weight of the driver and cultivator on the axle.

Cultivator wheels should be high and provided with wide tires.

Wheel boxes.—A notable improvement is found in the closing of the ends of the wheel boxes, making it possible to keep the bearings well lubricated.

The **spread arch** is a device to cause the gangs to swing in unison, and should be made adjustable in width.

Hitch.—It is a great advantage to have the height of hitch adjustable to horses of various sizes.

FIG. 67—A COMBINED WALKING AND RIDING SIX-SHOVEL CULTIVATOR WITH STRADDLE SEAT AND TREADLE GUIDE. THE HANDLES TO BE USED WHEN WALKING ARE NOT ATTACHED

Treadle guide.—Upon many cultivators a device has been added to guide the gangs as a whole by foot levers.

Such a device is called a treadle guide, and is often a very desirable feature.

Pivotal wheels are a scheme for guiding cultivators. The wheels may be connected to a treadle device or to a lever worked by the hands. This plan permits of an easy control of the cultivator.

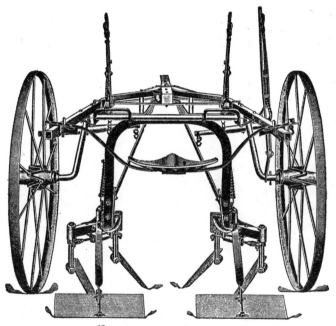

FIG. 68—A RIDING SURFACE CULTIVATOR

A walking, tongueless cultivator with four-shovel gangs is illustrated in Fig. 65. The tongueless offers one advantage in requiring less room for turning. It is essential that the team work very evenly to do good work. Fig. 66 illustrates a balance-frame six-shovel riding cultivator with a hammock seat. The wheels may be drawn back by a lever when the gangs are lifted in order to be more directly under the weight and prevent the tongue from flying up.

The combined cultivator, walking and riding, is illustrated in Fig. 67. This cultivator has a straddle seat and a balancing lever to adjust for the weights of different riders.

The surface, or the gopher, cultivator (Fig. 68) is used for surface cultivation. It is very effective in destroying

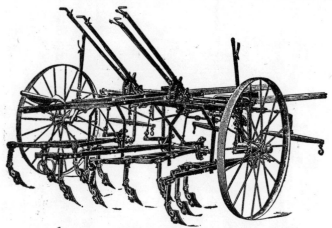

FIG. 69—A TWO-ROW CULTIVATOR, GUIDED WITH A LEVER

weeds when small, conserving the soil moisture, and does not prune the corn roots when working close to the corn.

The two-row cultivator is the latest production in the line of cultivators. It is a very useful tool where farm labor is scarce, and will do very creditable work for subsequent cultivations when the plants are of some height. Fig. 69 illustrates a cultivator of this type.

The disk cultivator illustrated in Fig. 70 is a tool which will move large quantities of dirt to or from the corn.

It is useful on this account for covering large weeds. Fig. 71 illustrates the eagle-claw gang, or the usual arrangement of shovels in the eight-shovel cultivator.

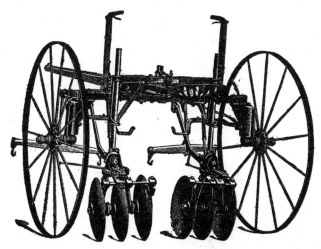

FIG. 70—A DISK CULTIVATOR

134. Listed corn cultivators.—For localities where the listing of corn is practiced, a cultivator has been

FIG. 71—AN EAGLE-CLAW FOUR-SHOVEL GANG

designed to follow the listed furrow for the first two cultivations. The machine is guided either by sled runners or roller wheels which run in the furrow. The shovel equipment varies between shovels and disks. The cultivator is made for one or two rows, and is a very successful tool.

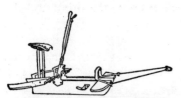

FIG. 72—A SIMPLE LISTED CORN CULTIVATOR. DISKS ARE OFTEN USED IN PLACE OF THE SCRAPERS. THE IMPLEMENT IS ALSO MADE TO CULTIVATE TWO ROWS AT A TIME

135. **Stalk cutter.**—An implement in general use in corn and cotton regions and which should be mentioned here is the stalk cutter. Its purpose is to cut cotton and corn stalks when left in the field into such lengths as not to interfere with the cultivation of the next crops. The implement primarily consists in a cylinder with five to nine radial knives. It is rolled over the stalks,

FIG. 73—A SINGLE-ROW STALK CUTTER

cutting them into short lengths. Stalk hooks are provided which gather the stalks in front of the cylinder. Two types are found upon the market, the spiral and the straight knife cutters. The spiral knife cutter carries practically all of the weight of the machine on the cylinder head while in operation, the side wheels being raised and the cylinder head brought in contact with the ground. Straight knife cutters have the cylinder head mounted in a frame, and when placed in operation are forced to the ground with spring pressure. The latter machine is much more pleasant to operate, as it rides more smoothly. Some cutters are equipped with reversible knives with two edges sharpened. A stalk cutter attachment is made for a cultivator carriage. The implement in general may be had as a single- or double-row machine.

CHAPTER VI

SEEDING MACHINERY

Seeders and Drills

136. Development.—Seeding by hand was practiced universally until the middle of the last century. Seed was either dropped in hills and covered with the hoe, or broadcasted and covered with a harrow or a similar implement. In fact, in certain localities in the United States hand dropping is practiced to some extent at the present time. Broadcasting seed by hand is practiced in many places.

A sort of drill plow was developed in Assyria long before the Christian era. Nothing definite is known of this tool, but it was evidently one of the crude plows of the time fitted with a hopper, from which the seed was led to the heel of the plow and drilled into the furrow. Just how the seed was fed into the tube we do not know. The Chinese claim the use of a similar tool 3,000 or 4,000 years ago.

Jethro Tull was perhaps the first to develop an implement which in any way resembles our modern drill. In 1731 he published a work entitled "Horse Hoeing Husbandry," in which he set forth arguments to the effect that grain should not be broadcasted, but should be drilled in rows and cultivated. This is, in a measure, like the system promulgated by Campbell, and which bears his name. Tull designed a machine which would drill three rows of turnips or wheat at a time. He used a coulter as a furrow opener and planted seed at three different depths His reason for this was that if one seeding failed, the others coming up later would be sure to be successful. Tull, like many others who spent their lives in invention, died poor, but he was successful in developing a line of drills, horse-hoes, and cultivators.

American development.—The first patent granted to an American was that Eliakim Spooner in 1799. Nothing remains to tell us of the nature of this device. Many other patents followed the first, but none are worthy of mention until a patent was granted to J. Gibbons, of Adrian, Michigan, August 25, 1840. Gibbons's patent was upon the feeding cavities and a device for regulating the amount delivered. A year later he patented a cylindrical feeding roll with different-sized cavities.

M. and S. Pennock, of East Marlboro, Pennsylvania, obtained a patent March 12, 1841, for an improvement in cylindrical drills. The patent pertained to throwing in and out of gear each seeding cylinder, and also to throwing the machine in and out of gear while in operation. These men manufactured their drill and sold it in considerable quantities.

Following the patent issued to the Pennock brothers came a long list of patents upon "slide" and "force-feed" drills. Slide drills are distinguished from the others in that a slide is provided to vary the size of the opening through which the seed has to pass, and in this way the amount of seed sown is varied. Force-feed drills carry the seed from the seed box in cavities in the seed cylinder, in which the amount is varied either by varying the size of seed pockets or by varying the speed of the seed cylinder.

The first patent upon a force-feed grain drill was issued November 4, 1851, to N. Foster, G. Jessup, H. L. and C. P. Brown, and was the introduction of the term force feed. In 1854 the Brown brothers incorporated as the Empire Drill Company and established a factory at Shortsville, New York. In 1866 C. P. Brown devised and patented a modification which has been known ever since as the "single distributer." One of Brown's employees, in connection with a Mr. Beckford, removed to Macedonia, New York, and in 1867 took out several patents which presented the "double distributer." The double distributer was a seed wheel with a flange on each side, one with large cavities and the other with small to suit the different sizes of grain. This system was adopted by the Superior Drill Company, of Springfield, Ohio. In 1877 a patent was granted to J. P. Fulghum for a device for varying the length of the cavities of the seed cylinder, and thus varying the amount of seed drilled. This principle is now used by many manufacturers.

The first drills were provided with hoes, but later a shoe was found to be more satisfactory. Perhaps the shoe was introduced by Brown, who devised the shoe for corn planters.

137. Classification of seeders.

Broadcast seeders:
Hand, rotating distributer.
Wheelbarrow.
End-gate, rotating distributer.
Wheeled broadcast:
  Wide track. Narrow track.
  Agitator feed. Force feed.
Combination with cultivator.
Combination with disk harrow.

138. **The hand seeder** with rotating distributer consists of a star-shaped wheel which is given a rapid rotation either by gearing from a crank or by a bow, the string of which is given one wrap around the spindle of the

FIG. 74—A CRANK HAND SEEDER. SEEDERS OF THIS KIND ARE ALSO OPERATED WITH A BOW

distributing wheel. Fig. 74 shows a seeder of this order. A bag is provided with straps which may be carried from the shoulders and the distributing mechanism placed at the bottom. The use of this seeder is confined to small areas, and the uniformity of its distribution of the seed is not the best.

139. The wheelbarrow seeder is used to some extent for the sowing of grass seed, and seems to be the survivor of this type of seeder, which was at one time used exten-

FIG. 75—A WHEELBARROW SEEDER

sively in England. A vibrating rod passes underneath the box and by stirring causes the seed to flow out of the openings on the under side of the seed box.

140. The end-gate seeder is provided with a rotating or whirling distributer much like the hand machine first described. Formerly nearly all of this type of machine

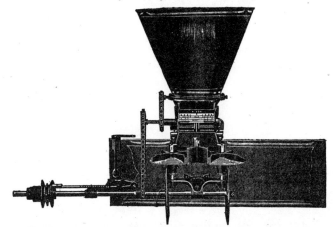

FIG. 76—AN END-GATE SEEDER WITH A FORCE FEED AND FRICTION GEARING. THIS MACHINE HAS TWO SEED DISTRIBUTERS

had only one distributer, but now the better makes are provided with two and a force-feed device to convey the seed to the distributer. Power to operate the seeder is obtained from a sprocket bolted to one wheel of the wagon on which the seeder is mounted, and transmitted to the seeder with a chain. The distributer is geared either by bevel or friction gears. It is stated that the friction gear relieves the strain on the machine when starting, and also runs noiselessly. The bevel gear drive

FIG. 77—AN AGITATOR-FEED BROADCAST SEEDER WITH CULTIVATOR COVER-ING SHOVELS. THIS IS A WIDE-TRACK MACHINE

is more durable and is recommended as being preferable by manufacturers who manufacture both styles of gears.

The same criticism may be made of this machine as of the hand machine. The distribution of the seed is not the best, and great accuracy in seeding is not possible. As the seeder is high above the ground, the wind hinders the operation of the machine to such an extent as to prevent its use in anything but a light wind or calm. In order to secure greater accuracy, the seed in some makes is fed to the distributer by a force-feed device. A small seeder of this type has been arranged to be placed upon a cultivator to sow a strip of ground the width of the cultivator as the ground is cultivated. This seeder has not as yet reached an extended use.

141. Agitator feed.—A broadcast seeder is still upon the market not provided with a force feed, but having what is known as an agitator feed. This feed is composed of a series of adjustable seed holes or vents in the bottom of the hopper, and over each is an agitator or stirring wheel to keep the seed holes open and pass the seed to them. The agitator feed, although cheaper and more simple than others, is not so accurate as the force feed described later.

Fig. 77 illustrates a broadcast seeder with an agitator feed and cultivator gangs attached. This seeder is usually used without any covering device; however, it may be procured with the cultivator gangs or with a spring-tooth harrow attachment.

FIG. 78—A FORCE-FEED DEVICE. THE FEED IS VARIED BY EXPOSING MORE OR LESS OF THE FLUTED FEED SHELL

142. Force-feed seeders and drills derive their name from the manner in which the grain is carried from the seed box. A **feed shell** is provided which is attached to a revolving shaft receiving its motion from the main axle.

FIG. 79—ANOTHER TYPE OF FORCE FEED

Fig. 78 shows the most common force-feed device. In the fluted cylinder, the device illustrated, the feed is regulated by exposing more or less of the cylinder to the grain. The feed shell is also designed in other ways. The seed cells may be on the inside and without any means of regulating the size of the cell. The feed or the amount of seed is regulated by varying the speed of the shaft carrying the feed shells by gearing as shown in Fig. 80.

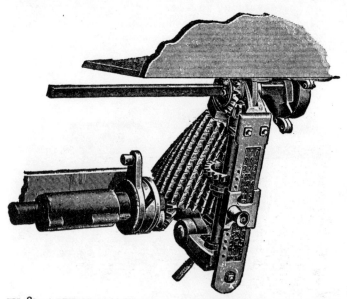

FIG. 80—A FEED-REGULATING DEVICE USED IN CONNECTION WITH A FORCE FEED SIMILAR TO THAT SHOWN IN FIG. 79

In order to handle successfully seeds of different size, the feed shell is made with two flanges with seed cells

FIG. 81—A FORCE-FEED BROADCAST SEEDER WITH NARROW-TRACK TRUCK

of different sizes in each. The cells best suited to the grain drilled are used, while the others are covered.

143. Width of track.—Broadcast seeders are now made

FIG. 82—A COMBINED DISK HARROW AND SEEDER. THIS MACHINE MAY ALSO BE SET TO DRILL FROM SEED SPOUTS AT THE REAR

with either wide or narrow track. Perhaps the wide track is the stronger construction and permits of higher wheels, but the narrow track permits of greater ease in turning and there is not the tendency to whip the horses' shoulders as with the wide track.

144. Combination seeders. — Broadcast seeders with cultivator and spring-tooth harrow attachments have been referred to. A popular tool now is the seeder attachment for the disk harrow. This attachment resembles very closely the force-feed broadcast seeder mounted above each of the harrow sections, and is operated by suitable sprocket wheels and chain from the main shaft of the disk. By the use of this tool two tools may be combined in one. The disk gangs, owing to their tendency to slip occasionally, do not make an entirely satisfactory drive. This is especially true in trashy ground. To surmount this difficulty, combination seeders are made with a follower wheel to drive the seeder.

DRILLS

Drills are provided with a force feed much like those used upon seeders, but are distinguished from each other in the type of furrow opener and covering devices used.

145. Classification of drills.

> Furrow openers:
>> Hoe.
>> Shoe.
>> Single-disk.
>> Double-disk.
>>> Covering devices:
>>>> Chains.
>>>> Press wheels.
>>>> Press wheel attachment.
>> Interchangeable disk and shoe drills.

146. The hoe drill was the first to be developed, and it is not difficult to see why this should be. The hoes are provided with break pins or spring trips in order that they may not be broken when striking an obstruction. These trip devices resemble very much those used upon cultivators. The hoe drill has good penetration, but clogs badly with trash. It is used extensively as a five-hoe drill for drilling in corn ground between rows of standing corn.

147. The shoe drill came into use about 1885 and has many advantages over the hoe drill. In fact, it was used almost entirely until the more recent development in the nature of the disk drill. Fig. 83 illustrates a shoe drill with high press wheels. The shoes are pressed into the ground with either flat or coil springs, which permit an independent action and prevent to a certain extent clogging with trash. It is claimed that flat springs do not tire as readily as coil springs, but coil springs seem to be almost universally used.

FIG. 83—A LOW-DOWN PRESS DRILL WITH SHOE FURROW OPENERS

148. Disk drills are the more recent development and consist of two classes: those with single- and double-disk furrow openers. In the single-disk type the disk is formed much like those used on disk harrows. Some form of heel or auxiliary shoe is provided to insert the grain in the bottom of the furrow made. It is desirable that the passage for the seed be so arranged that there can be but little chance for it to become clogged with dirt. The furrow opener that allows the seed to come into direct contact with the disk is not to be advised, but an inclosed boot should be provided to lead the seed into the bottom of the furrow. Some ingenuity is displayed by different makers in securing the desired results in this respect. In some drills the grain is led through the center of the disk. The single-disk may be given some

FIG. 84—A STANDARD SINGLE-DISK DRILL WITH A PRESS-WHEEL ATTACH-MENT. THE STEEL RIBBON SEED TUBES ARE ALSO SHOWN

suction, and therefore has more penetration than any other form of disk opener, fitting it especially for hard

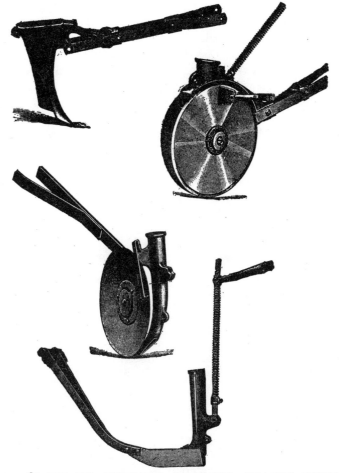

FIG. 85—THE HOE, DOUBLE-DISK, SINGLE-DISK, AND SHOE FURROW OPENERS USED ON DRILLS. THESE ARE OFTEN MADE INTERCHANGEABLE

and trashy ground. The single disk has one objection, and that is that it tends to make the ground uneven, since the soil is thrown in only one direction.

The double-disk furrow opener has two disks, or really coulters, as they are flat and their action is much like that of the shoe. One disk usually precedes the other by a short distance. The double-disk has not the penetration of the single-disk, but will not ridge the ground as the single-disk does. They often have another bad feature in that they allow dry dirt to fall on the seed, and hence prevent early germination. The single-disk drill does more to improve the tilth of the ground than any other furrow opener. The fact that a slight ridge is left in the center of the furrow with the double-disk is considered by some an advantage, as the seed is better distributed; in fact, two rows are planted instead of one.

149. Interchangeable parts.—Most manufacturers now design their drills in such a way that any one of the various styles of furrow openers may be used. Fig. 85 shows furrow openers which may be used on the same drill.

150. Press wheels.—Not a few years ago drills were equipped to a large extent with press wheels, but now they are not so popular. The press wheel, when sufficient pressure can be applied, is evidently a very good thing, as the earth is compacted around the seed and the moisture is drawn up to the seed, causing early germination. The pressure upon each press wheel must necessarily be very small, as most of the weight of the drill is required to force the furrow openers into the ground, and the balance is to be divided over a number of press wheels. It is not an uncommon thing to see an old drill running with some of the press wheels entirely off of the ground. Drills have been made in two distinct types, one known as the standard drill with the large wheels at the end of the seed box and equipped with small press wheels, and another where large press wheels were used and the large wheels at the end of the seed box dispensed with, which is spoken of as a **low-down** drill.

151. Press-wheel attachment.—In order to make their machine become more universal, manufacturers have provided press-wheel attachments for those who wish them, and they are detachable and do not interfere with the use of the drill whether with or without them. It is to be mentioned here that the drill has many conditions to meet, and a drill which will do satisfactory work in one section may not in another. Thus in a wheat territory, where the ground is not plowed every year a drill with great penetration is needed. In other sections where the ground is carefully prepared this particular feature is not so important. Press-wheel attachments are a nuisance in turning, and it is out of the question to back the machine.

152. Covering chains.—Chains are often provided to follow after the furrow openers, and their sole purpose is to insure a covering of the grain.

Formerly the grain tube or the spouts which convey the grain to the furrow opener were made of rubber, but the best used at the present time are made either of steel wire, or, still better, steel ribbon.

153. Disk drills.—Indications point toward the displacement of all forms of furrow openers by the single-disk opener. The single-disk will meet nearly all of the many conditions to be encountered. The double-disk is not much better in many respects than the shoe. The single-disk has good penetration, and besides is especially well adapted to cut its way through trash. Against it stand two objections: One is that there is a tendency for it to clog when the ground is wet, and the other is its weak point, the bearing. With the shoe drill, the wear is upon the shoe itself, but with the disk there is a spindle, and being so close to the surface of the soil, it is in a bad place to keep free from dirt and to lubricate. The bearings in use consist almost universally of chilled iron. Wood has proved itself to be especially well adapted for

FIG. 86—A STANDARD SINGLE-DISK DRILL WITH COVERING CHAINS

a place of this kind, but does not seem to be used. At any rate, in the purchase of a drill a close inspection should be made of the bearings to see that they are so designed as to give a large wearing surface, to be as nearly as possible dust proof, and to be provided with the proper kind of oil cups or other device for oiling.

154. Distance between furrow openers.—Drills are usually made 5, 6, or 7 inches between furrow openers. Perhaps 6 inches is the width generally used. They are placed 14 to 16 inches or more apart in the Campbell system, and then the grain cultivated during the growing season. It is thought desirable by some to have a slight ridge between the rows in order to hold the snow and to protect the young plant seeded in the fall from being affected so much by heaving. The action of the wind is to wear the ridges down, and in this way tend to cultivate the plants.

155. Horse lift.—The gangs of drills are very heavy and somewhat difficult to handle with levers, the levers being called upon to force the furrow openers into the ground while at work. To assist in this an automatic horse lift is provided on the larger drills.

156. Footboard.—To replace the seat a footboard is often placed on the drill. The operator in this case rides standing and is in a convenient position to dismount.

157. Grass-seed attachment.—The feed shell arranged for drilling the larger field grains does not have the refinement to drill grass seed with accuracy. It is often desired to drill the grass seed at the same time as the grain, and good results cannot be had by mixing and drilling together. The grass-seed attachment does not differ much from other devices except in size. Grass-seed attachments are often poorly constructed and become so open as to prevent their use after a few years' service.

158. Fertilizer attachment.—Practically all drill manufacturers can now furnish their machines with an attachment for drilling commercial fertilizer at the time of seeding. The fertilizer is usually fed by means of a plain rotating disk, which carries the fertilizer out from under the box. The seed mechanism will not work with fertilizer, as there is a great tendency to corrode on the part of some of the fertilizers.

159. The five-hoe or disk drill.—This tool is used for

putting fall grain in corn ground while the corn is standing. The disk drill has been displacing the hoe drill because it does not clog as easily with corn leaves. Fig. 87 shows a five-disk drill with a footboard so arranged that the operator may ride when it is necessary to add his weight to secure greater penetration.

160. Construction.—In purchasing a drill it might be well to investigate the construction. The implement, be-

FIG. 87—A FIVE-DISK DRILL FOR DRILLING BETWEEN CORN ROWS. THE CENTER FURROW OPENER IS A DOUBLE DISK

cause it is so heavy and often wide, should be provided with a strong frame. Angle bars or either round or square pipes are used to make the main frame. The frames are often provided with truss rods in order to stiffen them as much as possible. Some of the heavier drills are now made with tongue trucks much like disk harrows referred to in a preceding chapter. They are a very satisfactory addition.

161. Draft of drills.—Drills are not as a rule light of draft for the number of horses used. The following results are given from experiments made at the Iowa experiment station:

Drill	Kind of Disk	Distance Apart at Drill Rows	No. of Disks	Distance covered in Feet	Total Draft in Pounds	Draft per Foot
No. 4.	Double......	8″	10	6.7	450	67.1
No. 5.	Single.......	8″	10	6.7	460	68.6

Neither of the above drills was provided with any form of covering device other than chains. It is to be noted from the above tests that the single-disk drill requires more power than the double-disk in pulverizing the ground, but the difference is small.

162. Calibration.—The scales or gages placed upon drills and seeders to indicate the amount of seed drilled per acre are not as a rule to be depended upon for great accuracy. If they are correct at first, there is a tendency for them to become inaccurate as the drill becomes old. The operator should make calculations of the ground drilled and the amount of grain used, and in this way check the scale of the drill. Drills calibrated have shown the scale to be in error as much as 25 per cent.

163. Clean seed.—The drill is displacing, to a large extent, the broadcast seeder because the farmer desires to place all of the seed in the ground and at the proper depth. With the broadcast seeder, where various methods of covering of the seed are resorted to, the seed cannot be covered a uniform depth. Practically all fall seeding is now done with drills, and the broadcasting is used for the seeding of spring grains alone. Experiments at the Ohio, Indiana, North and South Dakota stations give, without an exception, better results from drilling, the increased yields for the drilling being from 2 to 5 bushels. In order to have a drill do its best work, great stress should be laid upon the fact that all grain should be clean and especially free from short lengths of weed stems, which are often found in grain as it comes from the threshing machine. These stems or pieces of straw may lodge in the feedway and prevent the grain from getting into the seed wheel.

CORN PLANTERS

164. Development.—Corn planters are strictly an American invention. This is not strange, for corn, or maize, is peculiarly an American crop. The development of the planter has also been recent; not much over 50 years have elapsed since the planter has been made a success. The Indians were the first to cultivate corn, but they never had anything but the most primitive of tools. Until the development of the horse machine, corn was almost universally planted and covered by means of the hoe, and in localities where a very limited amount of corn is grown the method is followed to-day.

The first machines used for seeding were universal in the respect that they were used for the smaller grains as well as corn. Perhaps the first patent granted on what may be styled a corn planter was given March 12, 1839, to D. S. Rockwell. In this planter may be seen in a somewhat primitive form some of the features of the modern planter. The furrow openers were vertical shovels, and the planter was supported in front and in the rear with wheels with the dimensions of rollers. The corn was dropped by means of a slide underneath the box. The jointed frame was patented by G. Mott Miller in 1843. George W. Brown, of Galesburg, Illinois, devoted much of his time to the development of the corn planter and secured patents on many features. To Brown's efforts is credited the shoe furrow opener, the rotary drop, and a method of operating the drop by hand. A patent on a marker was granted to E. McCormick in 1855 as a device projecting from the end of the axle. The present marker was set forth in a patent secured by Jarvis Case, of Lafayette, Indiana, in 1857. In about 1892 the Dooley brothers, of Moline, Illinois, brought out the edge-selection drop used extensively on the more recent planters.

165. Development of the check rower.—It seems that all the early planters were automatic, in that an operator was not needed to work the dropping mechanism. In 1851 a patent was granted to E. Corey, of Jerseyville, Illinois, for a device to mark the point where the corn was planted, and this device led to the use of a marker in laying off fields and putting the hills of corn in check. Brown's patent previously referred to was the first patent to cover the hand-dropping idea. M. Robbins, of Cincinnati, patented in 1857 a checking device for a one-horse drill using a jointed rod and chain provided with buttons for a line. The check rower was developed to a practical device by the Haworth brothers. The Haworth was for a long time the standard machine. The check wire in this implement was made to travel across the machine. Among the first of the side-drop check rowers was the Avery, which became at one time very popular. Recent changes in check rowers have been confined to reducing the amount of work done by the machine.

166. Hand planters have never come into any extended use, as they are not any great improvement over the hoe. This planter is made now much like it was years ago. Fig. 88 shows the common style and is used to some extent in replanting. A slide extends from one handle to the other and passes under the small seed box. When the slide is under the box a hole of the proper size is filled with the desired number of grains. When the handles are opened so as to close the points the hill of corn is drawn from under the seed box and allowed to fall to the point. There are modifications of this hand planter in which a plate is used

FIG. 88—A HAND CORN PLANTER. THE CORN IS DRAWN FROM UNDER THE SEED BOX BY A SLIDE UPON CLOSING AND OPENING THE HANDLES

and made to revolve by pawls which act by opening and closing the planter.

167. The modern planter.—Although most planters are called upon to do about the same work, they differ much in construction. The essentials of a good, successful

FIG. 89—A MODERN CORN PLANTER WITH LONG CURVED FURROW OPENERS, VERTICAL CHECK HEAD, AND OPEN WHEELS

planter have been set forth as follows: (1) It must be accurate in dropping at all times; (2) plant at a uniform depth; (3) cover the seed properly; (4) convenient and durable; and (5) simple in construction.

168. Drops.—The early planters had slides or plates in which holes or seed cells were provided which were large enough to hold a sufficient number of kernels to make an entire hill of corn. Planters are constructed in this manner and offer some advantages in dropping uneven seed. This style of drop is known as the **full-hill drop.**

The **cumulative drop** was the result of an effort to raise the accuracy of dropping. In the cumulative drop the grains are counted out separately (a seed cell being provided in the seed plate for each kernel) until a hill is formed, the theory of the accuracy being that there is less chance for one less or more kernels when the cell is nearly

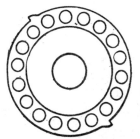

FIG. 90—THE ROUND-HOLE
SEED PLATE

FIG. 91—THE EDGE-SELECTION
PLATE

the size of each kernel, while in the larger cell three small kernels could easily make room for the fourth.

169. Plates.—The **round-hole** plate is a flat plate with round holes for seed cells; hence the name. The round-hole plate may belong to a full-hill or a cumulative drop planter.

The **edge-selection** or edge drop plate has deep narrow cells arranged on its outer edge, in which the corn kernel is received on its edge (Fig. 91). The arguments advanced in favor of this plan are that the corn kernel is more uniform in thickness than any other dimension, and owing to the depth of the cells is not so apt to be dislodged by the so-called cut-off. The majority of planter manufacturers within the past few years have brought out an edge-selection drop-plate planter and claimed great accuracy for it. Varieties of corn differ very much in the width of kernel, and for this reason provision has been made by at least one manufacturer to vary the depth of the edge-selection cell by substituting grooved bottoms to the seed box over which the plate travels. A device is provided with the flat plate for the same purpose. The outside edge of the cell is made open, into which a spring fits, excluding all but one kernel.

170. **Plate movement.**—Plates are made to revolve in a horizontal plane, and also in a vertical plane. To plates in these positions the names of **horizontal plate** and **vertical plate** are given, respectively.

The **intermittent plate movement** is one where the plate is revolved until a hill is counted out, and then remains at rest until put in motion for another hill by the check wire. The movement may belong to full-hill or cumulative drops. The argument is set forth that the seed cells are filled to better advantage by this intermittent motion; the starting and stopping will shake the corn into the cells. To cause the seed cells to fill more perfectly, the kernels are prearranged by the corrugations and the slope of the seed-box bottom.

In the **continuous plate movement** the plates are driven from the main axle usually by a chain and sprockets. While the plates travel continuously, the size of the hill is determined by a valve movement which opens and closes the outlet from the seed plate. To produce this movement, two clutches with double cam attachments, one at each hopper, are used. At each trip of the planter the dog on the clutch is thrown out, and it turns through one-half revolution, allowing one cam to pass; at the same time the arm of the valve glides over the cam and opens the outlet to the hopper, which allows the corn to drop from

each cell until the cam passes and the arm drops, closing
the valve. Thus the length of this cam determines the
length of time the valve is open, thereby controlling the
number of kernels in the hill. Several lengths of cams
are furnished with each planter. It is claimed in opposi-
tion to the claim set forth for the intermittent movement
that the cells are more apt to be filled, for they are in continuous motion and travel a greater distance under the
corn.

171. The clutch.—In the early planters the plate was
driven entirely by the check wire. With each button the plate

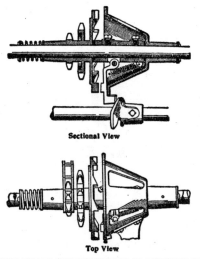

Sectional View

Top View

FIG. 92—THE SEED-SHAFT CLUTCH WHICH IS THROWN IN GEAR BY THE
CHECK WIRE. THE POWER TO DRIVE THE SEED SHAFT THEN
COMES FROM THE MAIN AXLE, NOT THE CHECK WIRE

was moved just far enough to deposit one hill in the seed
tube. When the cumulative drop was developed, a means
had to be provided to rotate the plate long enough to
count out the hill. To arrange for this, the button was

made to throw a clutch which put the dropper shaft in connection with a chain drive from the main axle. This clutch remained in gear for one revolution of the shaft, which is equivalent to one-fourth revolution of the seed plate. The one-fourth of the seed plate was arranged with enough seed cells to count out one hill. This clutch may be made to operate a valve which will permit a suffi-

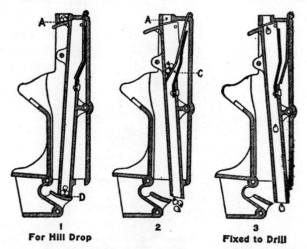

1   **2**   **3**
For Hill Drop   **Fixed to Drill**

FIG. 93—A TRIPLE VALVE MECHANISM SHOWING HOW THE CORN IS RE-
LEASED AT THE HEEL OF THE FURROW OPENER

cient number of kernels to leave the plate to make a hill as described above. The clutch has relieved the check wire of a large portion of its work. It is only required to put the clutch in gear and to open the valves in the shank. The clutch is one of the vital parts of the planter, and is often the first part to wear out and give trouble. Fig. 92 illustrates a planter clutch.

172. Valves are divided into three classes: single, double and triple valves. The single valve is placed in the heel of the furrow opener. The corn may be either caught here a single grain at a time or a full hill at a time. When the check wire throws the valve open to let a hill out, it closes in time to catch the next hill.

With the double valve, the hill is caught twice in its transit from the seed box to the ground. Fig. 93 shows one style of triple-valve arrangement.

The lower valves are made quite close to the ground and arranged to discharge backward and downward

FIG. 94—THE STUB-RUNNER FURROW OPENER

into the furrow to overcome the tendency to carry the hill on and make uneven checking.

173. Furrow openers.—The **curved runner** is used on a large majority of planters as a furrow opener. It is easy to guide, but will not penetrate trash or hard soil as well as some others. The curved runner is illustrated in Fig. 89.

The **stub runner** has good penetration and will hook under trash and let it drag to one side out of the way.

There is less tendency for the stub runner to ride over trash than the curved runner. Fig. 94 shows a stub runner. The stub runner cannot be used in stony or stumpy land.

The **single-disk furrow opener** has good penetration and is desired in some localities for that reason. It is also better adapted to trashy ground, the disks cutting their way through. The disks may or may not reduce draft;

FIG. 95—THE SINGLE-DISK FURROW OPENER

at any rate, the planter is not a heavy-draft implement. Penetration is not often needed; more often the planter has a tendency to run too deep. The single-disk planter throws the soil out one way, and it is difficult for the wheels to cover the seed. The disk has a bearing to wear out, which the runner has not.

The **double disk** cuts through trash to good advantage, but does not have the penetration of the single disk. It has two bearings to wear out to each furrow opener. It is claimed dry dirt falls in behind the disks on the corn, preventing early germination. All disk planters are very hard to guide. They do not follow the team well.

FIG. 96—A CORN PLANTER WITH DOUBLE-DISK FURROW OPENERS, OPEN WHEELS, AND HORIZONTAL CHECK HEADS

174. Planter wheels may be had in almost any height, from very low wheels to those high enough to straddle listed corn ridges. The tire may be **flat, concave,** or **open** (Fig. 97).

The **flat wheel** is not used to any extent on planters to-day, but is offered for sale by most manufacturers. It does not draw the soil well over the corn, but leaves this hard and smooth to bake in the sun, and gives the water a smooth course to follow after heavy rains.

The **concave wheel** gathers the soil better than the flat wheel, but leaves the surface smooth.

The **open wheel** is now used to a larger extent than any other type. It has good gather, covering the corn

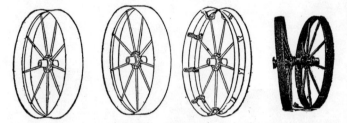

FIG. 97—CORN-PLANTER WHEELS WITH CONCAVE, FLAT, AND OPEN TIRE; ALSO THE DOUBLE WHEEL

well; the ground has no tendency to bake over the corn, and the water during rains is carried to one side of the track.

The **double wheel** consists in two wheels instead of one to cover the corn, and may be set with more or less gather, thus being able to cover the corn under all conditions.

175. Fertilizer attachment.—In some localities it is necessary to use fertilizer to secure an early and quick growth of corn. An attachment is made to drop fertilizer for each hill, and careful adjustment must be made to drop the fertilizer the right distance from the hill. If too far away, it will not give immediate benefits, and if placed too close, will rot the corn. This adjustment is difficult owing to the difference in speeds at which planters are operated.

176. Marker.—Markers are made in two styles, the **sliding** and the **disk**. The disk has proved to be a very satisfactory marker.

177. Wire reel.—Two types of check wire reels have been developed: one to reel by friction contact to the planter wheel, and one to reel under the seat with a chain to the main axle, using a friction clutch on the spool. It is claimed to be desirable to wind the wire on a solid, smooth drum rather than on a reel, as the former kinks the wire less.

178. Conveniences.—In making a purchase of a planter it is well to have in mind the conveniences which may be had, as well as the matter of strength, durability, and accuracy. Convenience in turning and reeling the wire is first to be considered. Another advantage offered by some planters over others is in the convenience of changing plates. It is very handy to have a seed box which may be tipped over and emptied without picking the seed out by hand.

The planter should have an adjustable tongue by which the front may be kept level. Unless the planter front is level, an accurate check cannot be obtained if the heel of the furrow opener is too far ahead or too far to the rear. It is impossible to get an even check if the planter front is not carried level.

It is desirable to have the check-rower arms act independently of each other, as it relieves the wire of some work. Two types of check heads for check rowers are used, the vertical and horizontal, both seem to be equally satisfactory.

179. Draft of planters.—Draft tests gave the following results for the mean draft of two styles of planters:

Planter with open wheels................212 pounds
Planter with double wheels.............237 pounds

180. Calibration of planters.—It is an undisputed fact that high accuracy cannot be secured with any planter unless the corn be of uniform size and a seed plate chosen to suit the size of corn. Types of corn vary much in size of kernel, and one plate will not suit all types and varieties. Makers usually furnish several plates with their machines, and others may be secured if necessary. It stands to reason that no planter can do good work unless these conditions are fulfilled. The planter should be calibrated and tested before taken to the field, if accuracy of work is desired.

181. Corn drills.—Although most planters may be set to drill corn, the corn drill remains a distinct tool and is

FIG. 98—THE SINGLE-ROW CORN DRILL

used to a large extent in certain localities of the country. Fig. 98 shows a single-row drill which differs but little from others except that an extra knife is provided in front of the seed tube. Various covering devices in the way of shovels and disks are provided. Drills are now made to take two rows, and even four, when made as an attachment to a grain drill.

182. Listers.—The use of the lister is confined to the semi-arid regions. It can be used in most of the corn-growing sections where the rainfall is not overabundant.

It is not adapted to fields that are extremely level, as water will collect in the ditches after rains and drown the corn while small. Neither can it be used in hilly localities, as the corn will in this case be washed out.

The lister is simply a double plow throwing a furrow both ways. The seedbed is prepared at the bottom of the furrow with a **subsoiler.** The planting may be done later or with an attached drill, which plants as the furrow is opened up. Thus plowing and planting are done at one operation. Fig. 99 shows one of the latest styles of walk-

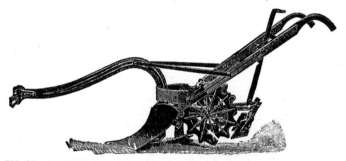

FIG. 99—A SINGLE-ROW WALKING LISTER WITH A CORN DRILL ATTACHED. DISKS ARE USED IN PLACE OF THE COVERING SHOVEL

ing listers with sprocket-wheel-drilling attachment. The drill attachment may be used independently as a drill. Fig. 100 is a representative three-wheel riding lister. Riding listers are also made without the furrow wheel, and when so made are termed sulky listers. Even the lister as a single-row machine has not been rapid enough for the Western farmer, and several makes of a two-row lister are to be found upon the market.

183. Loose-ground listers.—Listing of corn has some disadvantages. When listing is practiced, the soil is not all loosened, and when successive crops are grown in the same way an effect upon the yield is noticed. To gain the

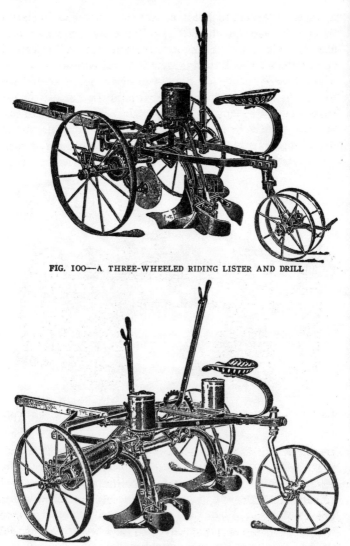

FIG. 100—A THREE-WHEELED RIDING LISTER AND DRILL

FIG. 101—A TWO-ROW LISTER

advantages of listing after plowing, the loose-ground lister has been developed. This tool is a two-row machine provided with disks to open the furrow, instead of right and left moldboards. Moldboards will not scour in loose ground, hence the use of disks. When the loose-ground lister is used, the ground must be plowed as for the planter, thus increasing the cost. The merits of the system consist in having the corn deeper to stand the drought better, and to be better braced to stand the high

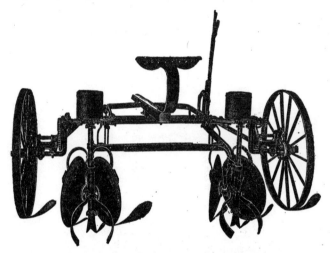

FIG. 102—A LOOSE-GROUND LISTER. DISK FURROW OPENERS MAY BE USED ON PLANTER FOR THE SAME PURPOSE

winds of the fall and not become "lodged." The fact that the corn is placed in a furrow makes it more easily tended because there is a large amount of soil to be moved toward the corn. In the moving of this dirt, any weeds are easily destroyed. Fig. 102 shows a loose-ground lister. Attachments are provided which may be placed upon corn planters to give the same results.

CHAPTER VII

HARVESTING MACHINERY

Agricultural machinery has done much for the agriculturist in enabling him to accomplish more in a given time, and to do it with less effort, than before its introduction. Although this is true of all agricultural machinery, it is especially true of harvesting machinery. By its use it has been estimated that the amount of labor required to produce a bushel of wheat has been reduced from 3 hours and 3 minutes to 10 minutes.

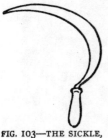

In this brief discussion harvesting machinery will be considered in its broadest sense and will include reapers, self-binders, headers, combined harvesters, and corn-harvesting machinery.

184. Development of hand tools.—From the oldest records that remain we find that the people of that early time were provided with crude hand tools for the reaping of grain. These primitive sickles, or reaping hooks, were made of flint and bronze, and are found among the remains left by the older nations. Upon the tombs at Thebes, in Egypt, are found pictures of slaves reaping. These pictures were made 1400 or 1500 B.C. The form of the Egyptian sickles varied somewhat, but consisted generally of a curved blade with a straight handle.

FIG. 103—THE SICKLE, AN EARLY HAND-REAPING TOOL

The scythe is a development from the sickle and differs from it in that the operator can use both hands instead of one. The Flemish people developed a tool known as the Hainault scythe. It has a wide blade 2 feet long, having a handle about 1 foot in length. The handle is bent at the upper end and is provided with a leather loop, into which the forefinger is inserted to aid in keeping the tool horizontal. The grain was gathered by a hook in the left hand. This tool was displaced later by the cradle.

Development in scythes has consisted in making the blade lighter, lengthening the handle, and adding fingers to collect the grain and to carry it to the end of the stroke. With the addition of the fingers, the tool was given a new name, that of the **cradle scythe,** or the **cradle.** And it was in this tool that the first American development took place. The colonists, when they settled in this country, probably brought with them all of the European types, and the American cradle was simply an improvement over the old country tools. The time of the intro-duction of the cradle has been fixed by Professor Brewer, of Yale, in an article written for

FIG. 104—THE AMERICAN CRADLE. THE TOOL USED FOR REAPING UNTIL AFTER THE MIDDLE OF THE NINETEENTH CENTURY

the Census Report of 1880, as somewhere between 1776 and the close of the eighteenth century.

The American cradle stands at the head of all hand tools devised for the reaping of grain. When it was once perfected, its use spread to all countries, with very little change in form. It has been displaced, it is true, by the horse reaper almost entirely; yet there are places in this country and abroad where conditions are such that reaping machines are impracticable and where the cradle has still a work to do. Again, there are parts of the world where the reaping machine has never been intro-duced and where the sickle and the cradle are the only tools used for reaping. It seems almost incredible that any people should be so backward as to be using at the present time these primitive tools, yet it is to be remembered that even the most advanced nations used them for centuries, and apparently did not think of anything in the way of improvement.

185. The first reaper.—History records several early attempts toward the invention of a machine for harvesting, but none reached a stage where they were practical until the eighteenth century. Pliny describes a machine used early in the first century which stripped the heads of grain from the stalk. The machine consisted of a box mounted upon two wheels, with teeth to engage the grain at the front end. It was pushed in front of an animal yoked behind it. The grain was raked into the box by the attendant as the machine was moved along. It is further stated that it was necessary to go over the same areas several times.

186. English development.—There were several attempts at the design of a reaping machine before 1806, but none were successful. They need not be considered in this discussion. It was in 1806 that Gladstone invented a machine which added many new ideas. In his machine the horse walked to the side of the grain, and hence the introduction of the side cut. It had a revolving cutter and a crude form of guard. It did, however, have a new idea in an inside and outside divider. The grain fell upon a platform and was cleared occasionally with a hand rake. As a whole, this machine was not successful.

In 1808 Mr. Salmon, of Woburn, invented the reciprocating cutter, which acted over a row of stationary blades. This machine combined reciprocating and advancing motion for the

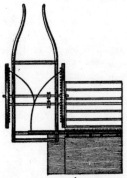

first time. The delivery of the grain was unique in the fact that a vertical rake actuated by a crank swept the grain from the platform upon which the grain fell after being cut.

In 1822, Henry Ogle, a schoolmaster of Remington, in connection with a mechanic by the name of Brown, designed and built a machine which is worthy of mention. The use of a reciprocating knife had been hinted at by Salmon, but Ogle made it a success. This machine also had the first reel used, and was provided with a dropper. Accounts are not specific, but it is thought that the operator for the first time rode upon a seat.

FIG. 105—OGLE'S REAPING MACHINE (ENGLAND, 1822)

The next machine was the most successful up to that time (1826). Patrick Bell, a minister of Cannyville, Forfarshire, has the honor of designing it. His machine had oscillating knives, each of which were about 15 inches long and about 4 inches broad at the back, where they were pivoted and worked over a similar set of knives underneath like so many pairs of shears. The rear ends of the movable blades were attached to an oscillating rod connected with a worm flange on a revolving shaft. It presented a new idea in having a canvas moving on rollers just behind the cutting mechanism, which carried the grain to one side and deposited it in a continuous swath. Bell also provided his machine with a reel and inside and outside dividers. His

FIG. 106—BELL'S REAPING MACHINE (ENGLAND, 1828)

machine marks the point when the development of the reaping machine was practically turned over to Americans. It never was very practical because it was constructed upon wrong principles, but nevertheless it was used in England for several years until replaced with machines built after the inventions of the Americans, Hussey and McCormick.

187. American development.—Beginning with the year 1803, a few patents were recorded before Hussey's first patent, which was granted December 31, 1833. These were not of any importance, since they did not add any new developments and were not practical. The only one which gave much encouragement was the invention of William Manning, of New Jersey, patented in 1831. Manning's machine had a grain divider and a sickle which were similar to those used later in the Hussey and McCormick machines.

It was in 1833 when Obed Hussey, of Baltimore, Maryland, was granted his patent which marks the beginning of a period of almost marvelous development. Though Cyrus B. McCormick was granted his first patent June 21, 1834, it is claimed that his machine was actually built and used before Hussey's, whose machine had the priority in the date of patents.

Hussey's first machine was indeed a very crude affair. It consisted of a frame carrying the gearing, with a wheel at each side and a platform at the rear. The cutter was attached to a pitman, which received its motion from a crank geared to the

FIG. 107—HUSSEY'S REAPING MACHINE (AMERICA, 1833)

main axle. The cutter worked in a series of fingers or guards, and perhaps approached the modern device much closer than any reaper had up to this time.

McCormick's machine was provided with a reel and an outside divider. The knife had an edge like a sickle and worked through wires which acted for the fingers or guards of Hussey's machine. The machine was of about 4½ feet cut and was drawn by one horse. The grain fell upon a platform and was raked to one side with a hand rake by a man walking.

Of the two machines, perhaps Hussey's had the more valuable improvement and it was nearer the device which proved to be successful later. Friends of both these men claim for them the honors for the first successful reaper. Hussey did not have the energy and the perseverance, and hence lost in the struggle for

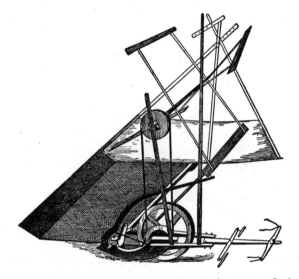

FIG. 108—M'CORMICK REAPING MACHINE (AMERICA, 1834)

supremacy which followed. At first the honors were evenly divided. In 1878 McCormick was elected a corresponding member of the French Academy of Sciences upon the ground of his "having done more for the cause of agriculture than any other living man."

Palmer and Williams, July 1, 1851, obtained a patent for a sweep rake which swept the platform at regular intervals, leaving the grain in bunches to be bound.

The next invention of importance was that of C. W. and

W. W. Marsh, of Illinois. A patent for this was granted August 17, 1858, and gave to the world the Marsh harvester. This carried two or more attendants, who received the grain from an elevator and bound it into sheaves. The two Marsh brothers, in connection with J. T. Hollister, organized a company which built 24 machines in 1864 and increased the output each year until in 1870 over 1,000 machines were built. This company was finally merged into the Deering Harvester Company.

George H. Spaulding invented and was granted a patent on the packer for the modern harvester, May 31, 1870. This invention was soon made use of by all manufacturers. John P. Appleby developed the packer and added a self-sizing device. He has also the honor of inventing the first successful twine knotter. The Appleby knotter, in a more or less modified form, is used on almost every machine to-day.

Jonathan Haines, of Illinois, patented, March 27, 1849, a machine for heading the grain and elevating it into wagons driven at the side of the machine.

In certain parts of the West, notably California, where conditions are such that grain will cure while standing in the field, a combined machine has been built which cuts, threshes, separates, and sacks the grain as it is drawn along either by horses or by a traction engine. The first combined machine was built in 1875 by D. C. Matteson. Benjamin Holt has done much to perfect the machine. The development of the grain harvester may be summarized as follows:

Gladstone was the first to have a side-cut machine.

Ogle added the reel and receiving platform.

Salmon gave the cutting mechanism, which was improved by Bell, Hussey, and McCormick.

To Rev. Patrick Bell must be given credit for the reel and side-delivery carrying device.

Obed Hussey gave that which is so important, the cutting apparatus.

For the automatic rake credit must be given to Palmer and Williams.

For a practical hand-binding machine the Marsh brothers should have the honor.

To Spaulding and Appleby the world is indebted for the sizing, packing, and tying mechanisms.

Jonathan Haines introduced the header.

Many other handy and important details have been added by a multitude of inventors, but all cannot be mentioned.

188. The self-rake reaper.—The modern self-rake resembles the early machine very much, and improvement has taken place only along the line of detail. The machine has a platform in the form of a quarter circle, to which the grain is reeled by the rakes, as well as removed to one side far enough to permit the machine to pass on the next round. The cutting mechanism is like that of the

FIG. 109—A MODERN SELF-RAKE REAPER

harvester. The machine is used to only a limited extent owing to the fact that the grain must be bound by hand. The reaper is preferred by some in the harvesting of certain crops, like buckwheat and peas. It is usually made in a 5-foot cut, and can be drawn by two horses, cutting six to eight acres a day.

MODERN HARVESTER OR BINDER

189. The modern self-binding harvester consists essentially of (1) a drive wheel in contact with the ground;

(2) gearing to distribute the power from the driver to the various parts; (3) the cutting mechanism of the serrated reciprocating knife, driven by a pitman from a

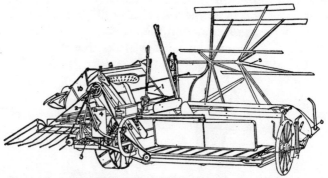

FIG. 110—A MODERN SELF-BINDING HARVESTER OR BINDER

crank, and guards or fingers to hold the grain while being cut; (4) a reel to gather the grain and cause it to fall in form on the platform; (5) an elevating system of endless

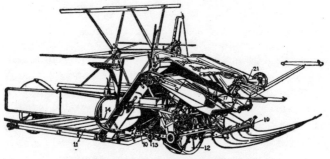

FIG. 111—ANOTHER MODERN HARVESTER

webs or canvases to carry the loose grain to the binder; and (6) a binder to form the loose grain into bundles and tie with twine.

Some of the more important features and individual parts will now be discussed in regard to construction and adjustment. Parts are numbered to correspond with numbers in Figs. 110 and 111.

190. Canvases (1) should be provided with tighteners by which they may be loosened when not in use. Tighteners also make it more convenient to put canvases on the machine. The elevator rollers should be driven from the top, thus placing the tight side next to the grain. The creeping of canvases is due to one of two things, either the canvases are not tight enough or the elevator frame is not square. If the elevator is not square, the slats will be torn from the canvases. This trouble may be overcome by measuring across the rollers diagonally or placing a carpenter's square in the corner between guide and roller, and adjusting. The method of adjustment varies with different makes, but the lower elevator is usually adjusted with a brace rod to the frame, and the upper elevator with a slot in the casting attaching the guide to the pipe frame.

191. Elevator chains (2).—Two kinds of chains are found in use, the steel chain and the malleable. The steel chain is claimed to be the most durable, but has the disadvantage of causing the sprocket teeth to cut away faster. This wear is often the greatest upon the driving sprocket, as it has the most work to do. It is thought that the steel chain is the more desirable chain to have.

192. The chain tightener (3).—The chain tightener may have a spring or slot adjustment. The spring adjustment is very handy and an even tension is maintained on the chain. The elevator chain should not be run with more tension than needed, as it produces wear and adds to the draft.

193. Twine box (4).—The location of the twine box is

the principal thing to be considered in order to secure the greatest convenience in watching the twine, and also in adding new balls.

194. Reel (5).—Convenience and strength are the principal things to be considered in a selection of a reel. It should have the greatest range of adjustment and permit this adjustment to be made easily. The making of a good bundle and the handling of lodged grain depend largely upon the manipulation of the reel. This may mean that the reel must be adjusted several times during a single round of a field.

The reel slats or fans should be adjusted to clear the dividers equally at each end, and also to travel parallel to the cutter bar.

195. Grain dividers (6).—It is an advantage to have the outside divider adjustable not only for different-sized grain, but also for making the machine narrow when mounted upon the transport trucks.

196. Grain wheel (7).—The weak point of the grain wheel is the bearing, and it is often necessary to replace the axle and boxings several times during the life of a machine. In order to prolong the life of the grain-wheel axle, it is made, by some manufacturers, with a roller bearing.

197. Elevators (8).—The elevator should extend well to the front of the platform in order that the grain may not be hindered in the least in starting upon its path up the elevator. The guides should be hollowed out slightly on the side next the grain, giving the canvases a chance to expand and not drag heavily upon the guides. It is also an advantage to have the lower end of the upper guide flexible in order that it may pass over extra large bunches of grain. The open elevator, permitting the handling of long grain, as rye, is now almost universally adopted.

The sprockets by which the elevator rollers are driven should always be in line. Adjustment may be made by sighting across their face.

198. Deck (9).—The steeper the deck, the better; but makers have made it rather flat in order to reduce the height of the machine. The deck should be well covered by the packers to prevent clogging.

199. Main frame (10).—Main frames are shipped either separate or fastened to the platform. In the latter case, if there is a joint, it is riveted and very seldom gives any trouble in becoming loose. In the first case, bolts must be used; but they do not give any trouble if care is used in assembling the binder.

200. Platform (11).—The platform is now universally provided with an iron bottom, which is more durable and smoother for the platform canvas to pass over. It is made of painted iron, and it might be improved if it should be made of galvanized iron, as it often rusts out before the machine is worn out.

201. Main wheel (12).—The main wheel is one of the parts which usually outwear the rest of the machine. The tendency is now to make the main wheel too small. The larger wheel is more desirable, as it carries the load better and is able to give a greater driving power. Main wheels have now attained a standard size of 34 and 36 inches in the side-cut machine. The steel wheel is now used almost universally, the wooden wheel and the wooden-rimmed wheel having gone out of use entirely. Three types of spokes are used: the hairpin, the spoke cast in the hub, and the spoke fastened to a flange of the hub with nuts. The main wheel shaft or axle should be provided with roller bearings, and also a convenient and sure method of oiling. The bolt in the lower part of the quadrant should always be in place. When the bolt is out it is possible to run the machine up too far and let the main axle start into the quadrants crosswise.

202. Main drive chain (13).—Two common types of drive chains are to be found upon the market: the all-malleable link and the malleable link with the steel pin. The latter is perhaps the more desirable, but not so handy for replacing broken links. The main chain should not pass too close to the tire of the main wheel, or it will clog with mud badly.

203. Cutter bar (14).—Two kinds of cutter bars are found, the Z bar and the angle bar. One seems to be as good as the other, but some little difference is to be found between the angle given to the guards, enabling some machines to cut closer to the ground than others.

204. Main drive shaft (15).—The main drive shaft should be given good clearance from the main wheel to prevent clogging. This shaft is now generally provided with roller bearings, and often self-aligning bearings, which prevent any possible chance for the shaft to bind and thus increase the friction.

205. Butter or adjuster (16).—The canvas butter has always been very satisfactory, except it was short-lived. Often it was the first part of the binder to be replaced. This led several makers to build an adjuster which had oscillating parts or board. The single board seems to be just as satisfactory, as the upper half of the two-board adjuster does very little good. The all-steel belt as now commonly used upon push binders is no doubt the most satisfactory butter made. It is durable and efficient, but not generally adopted, probably on account of its cost.

206. Packers (17).—The packers should practically cover the deck, reaching within an inch or so of the deck roller. This will prevent any tendency to clog in heavy grain.

The third packer is considered an advantage, but is not generally adopted.

207. Main gear (18).—Considerable difference is to be noticed in different binders in the size of the gearing used. It is true that many makers are not liberal enough with material in the construction of the main gear wheels.

208. Bundle carriers (19).—Two general types of bundle carriers are to be found in use. In one the fingers swing back when depositing the load, while in the other the carrier is simply tipped down at the rear and the load of bundles allowed to slide off. The swinging bundle carrier scatters the bundles quite badly. While the other does not have this fault, it does not work so well in hilly countries, because in going downhill the bundles refuse to slide from the carrier, and in going uphill they will not stay on the carrier.

209. Tension (20).—The roller tension, introduced a few years ago, is without doubt the best device of the kind

FIG. 112—A ROLLER TENSION FOR THE TWINE. A VERY SATISFACTORY DEVICE

yet invented. The twine will not be caught at knots, kinks will not be formed, and the tension is always even independent of the size of the twine.

The tension should not be used to produce tight bundles. It should be used only to keep the twine from playing out too fast.

210. Binder attachment (21).—The mechanism which ties the bundle is usually spoken of as the binder attachment. The first binder attachment depended upon a train of gear wheels to transmit the power to the needle and the knotter mechanism. At least one binder still retains this feature, while others have adopted the shaft and bevel gears, a chain and sprockets, or a lever in some form or other. Each binder, however, seems to be satisfactory in this particular. The levers have perhaps a disadvantage in that a very slight wear produces a marked effect upon the adjustment of the parts. The clutch is one of the important features of a binder attachment and perhaps demands of the expert more attention than any other one part of the binder. If the attachment stops before a bundle is made, even though it may be for but a short time, the action would indicate something to be wrong with the clutch. The binder attachment is driven directly from the crank shaft in some makes and in others by the elevator chain. The former method is to be preferred, as it relieves the elevator chain of part of its work.

211. Knotter (22).—The term knotter is applied to the knotter hook or the part on which the knot is produced, and also to the entire mechanism making the knot, including frame, knotter hook or bill, knotter pinion, knife, disk, gear, etc. (See Fig. 113.)

The knotter has been changed but little since it was first introduced by Appleby. The worm gears have to some extent been replaced by cam motion, which is more adjustable. Simplicity of parts may or may not be an advantage. An adjustable device to drive the twine disk, for instance, is often a great advantage. A stripper to carry the twine from the knotter hook has proved more reliable than to depend upon the twine being pulled from the knotter hook by the bundle.

212. Adjustment.—It seems impossible to take up the adjustment of the binder in the light of experting in this treatise. However, there are a few misadjustments of

FIG. 113—A TWINE DISK. A KNOTTER COMPLETE—A KNOTTER HOOK

common occurrence, and often resulting in loss of dollars to the user of the machine, which may be taken up here.

1. A loose main drive chain permits the chain to ride the teeth of the sprocket and slip down the teeth, giving the machine a jerky motion, as if some part was catching and stopping the machine. A dry or muddy chain aids in giving this effect.

2. If the slats are torn from canvases, the elevators are not square or the rollers are not parallel to each other. The method of putting elevators in square has been explained.

3. If the main gear cuts badly and wears rapidly, either the gears do not mesh properly or the elevator chain is too tight.

4. The knotter hook will not work properly unless smooth and free from rust. It can be polished with fine emery paper.

5. The binder attachment will not do its work prop-

erly unless timed. By this is meant the adjustment of each part so it will do its share at the proper time. Marks are placed on the teeth of gear wheels and sprockets to enable them to be properly timed. Some binders are timed in as many as five places.

6. The knotter pinion must fit to the tyer wheel, and there must not be any lost motion. The tyer wheel, or cam wheel, may be set up against the knotter pinion, but if worn the knotter pinion must be replaced. If the knotter hook does not turn far enough to close the finger on the twine, a knot will not be tied.

7. If the cord holder does not hold twine tight enough, the twine will be pulled out before the knot is made. It should require a force of about 40 pounds to pull the twine from the disk. Adjustment is made with the cord-holder spring.

8. If the disk does not move far enough, the knotter hook will grasp only one cord; hence a loose band with a knot on one end.

9. If the needle does not carry the twine far enough, the hook will grasp only one cord, and hence a loose band with a loose knot. The travel of the needle is adjusted by the length of the pitman. The needle may become bent, as it is made of malleable iron, but it will permit of being hammered back into form.

10. If the knife is dull, it may pull the twine from the hook before the knot is made.

11. The compress spring relieves the strain on the machine when the needle compresses the bundle. It should never be screwed down until dead in an effort to make larger bundles.

12. The bundle-sizer spring—not the tension or compress spring—should be used to make tight bundles.

13. Good oil should be used and all holes kept open. In setting up new machines, kerosene should be used to loosen up the paint.

14. Any difficulty must be traced to its source, and adjustment should not be made haphazard in hope of finding the trouble.

213. The transport truck.—When it is necessary to move the binder from place to place, it is mounted upon transport trucks, which facilitate its transportation through gates and over bridges. The trucks are set under the machine by raising the machine to its maximum height and then lowering it to the trucks. The tongue is then removed and attached at the end of the platform beside or through the grain wheel. Some transports are more handy to attach than others.

214. The tongue truck.—Owing to the weight on the tongue and the fact that the team cannot be well placed directly in front of the machine to prevent side draft, the use of tongue trucks has become popular, especially on the wide-cut machine. Their use is to be commended, for not only is the work made easier for the horses, but it permits four horses to be hitched abreast.

215. Width of cut.—Binders vary in the width of cut or swath from 5 to 8 feet. The 6-foot machine is the common size to be used with three horses, the harvesting of 10 to 15 acres being an average day's work. The 7- and 8-foot machines are used in localities growing lighter crops and require four horses.

216. Draft of binders.—The following results were obtained during the season of 1906 at Iowa State College from testing a McCormick and a Deering binder cutting oats. The ground in both cases was dry and firm.

McCormick 6-foot: Average of three tests....316 pounds
Deering   6-foot: "   "   "   ....312 pounds

217. The header is a machine arranged to cut the standing grain very high, leaving practically all of the straw in the field. The cutting and reeling mechanisms of the header are much like those of the harvester, but the machine differs decidedly in the manner of hitching the teams for propelling it. It is pushed ahead of the horses and guided from the rear by a rudder wheel. The headed grain is carried by canvases up an elevator and deposited in a wagon with a large box drawn along beside the machine. The header usually cuts a wide swath from 10

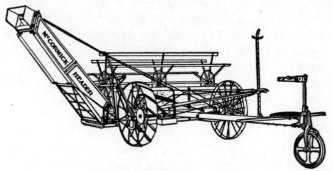

FIG. 114—THE MODERN HEADER

to 20 feet, and requires 4 to 6 horses to operate it. With it, 20 to 40 acres may be harvested in a day. An attachment is sometimes placed upon the header to bind the cut grain into bundles, in which case the grain is cut lower. This attachment must necessarily be very highly geared, but does very satisfactory work. A machine with a binder attachment is called a **header binder**.

218. The combined harvester and thresher is a threshing machine with a harvesting mechanism at the side which conveys the headed grain from a wide swath directly to the thresher cylinder. The cutting and elevating machinery is much like that of the header, and the threshing machine is of the usual type. It is to be mentioned that this machine can be used only where the grain will cure while standing in the field, and where the climate provides a dry season for the harvest. These machines have an enormous capacity, harvesting and threshing up to 100 acres or to 2,500 bushels of grain a day. The swath varies from 18 to 40 feet. The power

FIG. 115—THE COMBINED HARVESTER AND THRESHER OPERATED BY STEAM POWER

may be furnished either by horses or a traction engine. From 24 to 36 horses or mules are required to furnish the power. All the horses or mules are under the control of a pair of leaders driven by lines. Following the leaders there are usually two sets of four, and the remainder of the animals are arranged in sets of six or eight. In this way one man is enabled to drive the entire team. At least three other men are required to operate the machine, one to have general supervision, one to tilt the cutter bar, and one to sew and dump the sacks when they accumulate in lots of six or eight. The largest machines are operated by steam power.

CORN HARVESTING MACHINERY

219. Development.—The corn binder has become in recent years a very important tool because farmers have begun to realize the true worth of the cornstalk as feed for live stock.

It has been stated by good authorities that 40 per cent of the feeding value of the corn lies in the leaves and stalks. To let all this go to waste is, to say the least, poor economy, but to handle the corn crop entirely by hand is so laborious that it was not until modern labor-saving tools were developed that the saving of the entire crop could be practiced. It is true that the ear and the stalk have been used for stock food from the earliest time, but the practice was always limited in the corn belt as long as hand methods prevailed.

The earliest tool used for cutting corn was the common hoe, and certainly must have been a very awkward tool. Later the sickle was made use of in topping the corn, a method by which the stalk was cut off above the ear after fertilization had taken place. Methods used in an early time for the building of shocks or stooks of corn would seem very crude to-day. Often a center pole was sunk into the ground and horizontal arms inserted in holes in it. Against this the corn was piled until a shock of sufficient size was formed, then the arms were withdrawn, finally the center pole. The whole was compressed and tied with a cornstalk band. Another method used to-day is to tie the tops of four hills, forming a saddle against which the corn is piled.

The corn knife was soon developed, and was first perhaps an old scythe blade provided with a handle. The manufactured corn knife can now be bought in a variety of shapes and with a choice of handles. One style of knife may be fastened to the boot, but does not seem to be very successful.

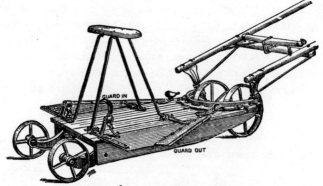

FIG. 116—A SLED CORN HARVESTER

D. M. Osborn & Co., as early as 1890, presented a corn harvester to the public. It cut the standing corn and elevated it into a wagon drawn beside the machine. The McCormick corn binder was soon to follow. It is a striking fact that the first McCormick machine was a machine pushed before the horses. In 1893 the Deering corn harvester was given a field

FIG. 117—THE VERTICAL CORN HARVESTER

trial, which was claimed to be very successful. To-day there are several corn harvesters upon the market.

220. Sled harvesters.—Many attempts were made following the introduction of the grain binder to build a corn harvester, but all resulted in failures. The sled harvester

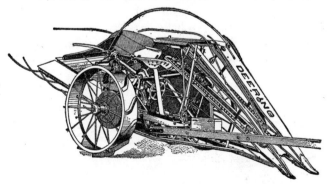

FIG. 118—THE HORIZONTAL CORN HARVESTER

was the first successful machine. It consists of a sled platform or a platform mounted upon small wheels, which carries knives at an angle to cut the corn as it is grasped by the operator, who rides on the platform. The machine is made for one horse, with the knives sloping back from the center, or for two horses, with the knives sloping from the outside to the center. This machine is cheap and has a much larger capacity than hand cutting. Heavy corn cannot well be handled, however, with a sled harvester.

221. Types of harvesters.—Corn harvesters may be divided into three classes, depending upon the position of the bundle while being bound. This may be either in a **vertical, inclined,** or a **horizontal** position (Figs. 117 and 118).

The vertical harvester seems to be the most popular, although the other types do very satisfactory work. Owing to the difference in the height of corn in various parts of the country, some makers provide two styles of harvesters, one for short corn and the other for tall.

The binder of the corn harvester resembles very closely the binder of the grain harvester. At first they were identical, but later it was found best to make the binder for the corn harvester a little heavier. The corn harvester should be provided with roller bearings and other conveniences of adjustment to be found upon the grain binder.

222. The stubble-cutter attachment consists of a knife attached to the corn harvester. It cuts the stubble close to the ground and makes further operations in the field more convenient. The attachment does not add much to the draft of the machine, and is surely a very useful device.

223. The corn shocker was one of the first machines

FIG. 119—A CORN SHOCKER. THE CORN SHOCK IS COLLECTED ON THE PLATFORM AND LIFTED TO THE GROUND AFTER BEING TIED

devised by the early inventors for the handling of the corn crop, but it was not presented to the public until after the introduction of the corn harvester or binder. It resembles the corn harvester in the construction of the dividers and the cutting mechanism. Fig. 119 illustrates the modern corn shocker. To the rear of the dividers a rotating table is placed with a center post. The corn is guided by fingers and angle irons to the center of the table. As additional stalks are cut they are added to the outside until a shock of proper size is formed. The machine is stopped and the shock tied with twine. By the aid of a windlass and crane the shock is lifted bodily from the table and dropped to the ground. When the tension on the lifting rope is slacked the arms which enabled the shock to be lifted are released by pawls, so they no longer remain in a horizontal position, but turn down as the center post is drawn from the shock.

The capacity of the corn shocker is only about one-half that of the corn harvester. It has the disadvantage that only small shocks can be made, which do not stand well and blow down easily. Another objection to its use is that the corn is more difficult to handle than when bound into bundles. There is, however, a saving of twine, and the work involved is not so laborious as that of shocking corn bundles by hand.

224. Loading devices.—The past few years have witnessed the introduction of several devices for loading corn fodder, hay, manure, etc. The machine usually consists of a crane or derrick with a horse lift by which a fork large enough to handle an entire shock is brought into action.

225. Corn pickers.—There have been many attempts to make a corn picker which would pick the ears from the standing stalk. For many years these attempts resulted in failures. However, the present scarcity of farm labor and the liberal prices paid for the picking of corn have again encouraged many inventors to spend time and money upon a machine of this kind. During the recent seasons several makes of corn pickers have been tried, with more or less success. Without any doubt, it is only a question of time until a practical machine may be had.

Two general types of corn pickers are to be found: the corn picker proper and the corn picker-husker. The former does not attempt to husk the ears, but simply to remove the ears from the stalk. However, in this opera-

FIG. 120—THE CORN PICKER-HUSKER

tion a large portion of the husks are removed from the ear. The remaining husks do not greatly interfere with the feeding or shelling of the corn.

The other type is provided with husking rolls, which remove the husks before the ears are elevated into a wagon drawn beside the machine.

CHAPTER VIII

HAYING MACHINERY

The introduction of modern haying machinery has wrought almost the same change in the harvesting of the hay crop as harvesting machinery has in the harvesting of the small-grain crop. The labor involved under present conditions in the cutting, curing, and storing of a ton of hay is but a small fraction of what it was under the old system of hand methods.

The hay crop ranks third in value among our crops. The addition of several new plants has greatly increased the value of the hay crop. This is especially true of alfalfa and brome grass, which have proved to be very valuable hay crops. The practice of curing grass for forage was in vogue before written history was begun. The first tools were as crude as possible. To-day we have a very complete line of hay tools for all conditions of work.

THE MOWER

226. The mower.—The development of the mower has been traced by M. F. Miller in the "Evolution of Harvesting Machinery,"* a bulletin published by the United States Department of Agriculture, and we are pleased to quote as follows:

"In the early development of the mower it was so intimately connected with the reaper that a little space should here be devoted to a short review of its history. Hussey's first machine was really a mower, and it was upon this principle that the mower was afterward built. Many of the early machines con-

* Bulletin No. 103, Office of Experiment Stations.

tained combinations of the mower and the reaper, and were used with a little adjustment to cut either grain or grass. A name that stands out prominently in the development of mowers is that of William F. Ketchum, who has sometimes been spoken

FIG. 121—KETCHUM'S MOWER (AMERICA, 1847)

of as the father of the mower trade, since he was the first to put mowers on the market as a type of machine distinct from the reaper. He took out several patents, but the one granted July 10, 1847, was of especial importance. The main features of this patent were the unobstructed space left between the driving wheel and the finger bar, with its support, and the remarkable simplicity of the machine. The cutter was an endless chain of knives, which never became successful, but which caused some excitement at the time. Ketchum afterward adopted the Hussey type of cutter and produced a very successful mower of the rigid-bar type. It was this machine that led the way in mower development and became the first really practical machine. . . .

"The first invention showing the feature of a flexible bar was that of Hazard Knowles, the machinist of the Patent Office at Washington. It showed many valuable features of a reaping machine also, but no patent was taken out. The patent granted to Cyrenus Wheeler, December 5, 1854, marks the division between the two types of machines. Wheeler was a practical man, and, like McCormick in the development of the reaper, suc-

ceeded in combining so many important features in his machines
as to give him a place as one of the foremost pioneers in the
development of the mower. The machine of 1854 was not a success as constructed, but the features of two drive wheels and
a cutter bar jointed to the main wheels were lasting. . . .

"On July 17, 1856, a patent was granted to Cornelius Aultman
and Lewis Miller containing principles that still exist in all
successful mowers. The first patent claimed 'connecting the
cutter bar to the machine by the double-rule joint or the double-
jointed coupling pin.' It was reissued to cover an arrangement
for holding up the bar while moving, and the combination of
ratchet-wheel pawl and spring. On May 4, 1858, Lewis Miller
took out a patent on a mower that combined the features of
the former machine with some new principles. It contained
all the elements of the successful modern two-wheeled machine,
and mower development since that time has been a perfecting
of this type. This machine was built under the name of the
'Buckeye,' and, with a substitution of metal for certain wooden
parts, and certain other improvements, it is in use to-day.
E. Ball, associated with this firm, also made valuable improvements in mowers. In 1856 a patent was granted to A. Kirby
covering improvements made by him a few years previous, and
his machines soon became popular. Others took up the manufacture of mowers at this early date, so that by 1860 the mower
had become a thoroughly practical machine, and was being
improved by various firms throughout the country. This improvement has gone on with the many makes of machines now
in existence, and to-day we have various forms, from the single
one-horse machine to the large two-horse type, with its long
cutter bar, running with as light a draft as the former clumsy
machine did with a cut but half as wide. As a result of this
development the amount of hay produced in the United States
has increased enormously, and to-day it stands as one of the
most important crops."

MODERN MOWERS

227. Types.—Modern mowing machines are of two
types, the side-cut mower and the direct-cut mower. The
cutter bar of the former is placed at one side of the drive

wheels or truck, while in the latter it is placed directly in front of the drivers. The mower consists essentially in (1) the cutting mechanism, comprising a reciprocating knife or sickle operated through guards or fingers and driven by a pitman from a crank, (2) driver wheels in contact with the ground, (3) gearing to give the crank proper speed, and (4) dividers to divide the cut grass from the standing.

228. The one-horse mower is usually a smaller size of the two-horse machine, fitted with shafts or thills instead of a tongue. It is made in sizes of 3½- or 4-foot cut, and is used principally in the mowing of lawns, parks, etc.

229. The two-horse mower is commonly made in 4½- and 5-foot cuts, although 6-, 7-, and 8-foot machines are

FIG. 122—A MODERN TWO-HORSE MOWER

manufactured. The latter are spoken of as wide-cut mowers and are usually of heavier construction than the standard machines (Fig. 122). From 8 to 15 acres is an average day's work with the 5- or 6-foot machines.

230. Mower frame.—Mower frames are usually made in one piece of cast iron. The openings for the axle and the shafting are cored out, but where the bearings are to be located enough extra material is provided for boring out to size. Roller bearings are usually provided for the main axle.

231. The crank shaft is usually provided with a plain bearing at the crank and a roller bearing at the pinion end. A ball bearing is provided at the end of the small bevel pinion to take the end thrust. It is not possible to use a ball or roller bearing at the crank end, due to the vibratory action of the shaft tending to wear the bearing out of round. This bearing is either provided with an adjustment or an interchangeable brass bushing to take up the wear. The crank should be well protected from the front and under sides. The crank and pitman motion seems to be the most satisfactory device to transmit a reciprocating motion to the knife. A wobble gear was tried a few years ago, but has been given up. A mower is manufactured with a pitman taking the motion from the face of the crank wheel instead of the side. It is not known how successful this machine is.

232. Main gears.—The driving gears should be liberal in size and always closed in such a way as to be protected from dust, and also to facilitate oiling. It might be an advantage in mowers as in some other machines to have the gears run in oil.

233. Wheels should be high and have a good width of tire. The common height is 32 inches, and 3½ and 4 inches the common width of tire. It is some advantage to have several pawls to engage the ratchet teeth in the wheels, because this feature, in connection with a clutch with several teeth for throwing the machine in and out of gear, will make the machine more positive in its action. That is, the sickle will start to move very shortly after the main wheels are set in motion. Mowers driven by large gear wheels in the drive wheels are more positive in their action and hence are preferred in foreign countries where very heavy swaths are to be cut.

234. The pitman in the mower corresponds to the connecting rod in an engine. Its function is to change circular motion into rectilinear motion, the reverse of the connecting rod. The crank pin and sickle should always be at right angles with each other, but this feature is not so essential when the pitman is connected to the sickle with a ball-and-socket joint.

Pitmans are made of wood and steel. Wood rods are the most reliable, because steel, due to the excessive vibration, becomes crystallized and weak. The steel pitman, however, may be so constructed as to be adjustable, and enables the operator to adjust the length until the knife acts equally over the guards at each end of the stroke. The pitman should be protected from being struck by any obstruction from the front.

235. The cutter bar is the cutting mechanism, exclusive of the sickle. It has a hinge coupling at one end and a divider and grass board at the other. The bar proper to which the guards are bolted should be stiff enough to prevent sagging. It is the practice in some machines to make the bar bowed down slightly and to straighten it by carrying the greater part of the weight at the hinge end, the weight of the bar itself causing it to straighten.

Some arrangement should be provided to take up the wear of the pins of the hinge joints in order that the cutter bar may be kept in line with the pitman.

236. Wearing plates.—Best mowers are now equipped with wearing plates where the sickle comes in contact with the cutter bar. They may be renewed at a small cost. The clips to hold the sickle in place are now made of malleable iron and are bolted in place to facilitate their replacement when worn. If slightly worn, they may be hammered down until the proper amount of play between the clip and the sickle is obtained. Under normal conditions, this is about 1/100 of an inch. In no case should it be so open as to permit grass to wedge under the clips, but at all times should hold the knife well upon the ledger plates so as to give the proper shearing action.

237. Mower guards are fitted with two kinds of ledger plates, one with a smooth edge and the other with a serrated edge. The serrated plate holds fine grasses to better advantage than the smooth ledger plate, and in this way aids with the cutting.

238. Shoes.—The cutter bar should be provided with an adjustable shoe at each end, by means of which the height of cut may be varied to some extent. A weed attachment is often provided which will enable the cutter bar to be raised 10 inches or more. A shoe is better than a small wheel at the outer end of the bar because the wheel will drop into small holes, while the runner will bridge them.

239. The grass board.—The purpose of the grass board and the grass stick is to rake the grass away from the edge of the swath to give a clean place for the inside shoe the next round. The grass board should be provided with a spring to make it more flexible and less apt to be broken in backing and turning.

240. Foot lifts.—Nearly all modern mowers are now provided with a foot lift, which enables the operator to lift the cutter bar over obstructions, and also makes easier work for the team by lifting the bar while turning. A spring is necessary to aid in the lifting.

Certain mowers, known as vertical lift mowers, permit the cutter bar to be lifted to a vertical position by a lever, to pass obstructions, and at the same time the mower is automatically thrown out of gear. When the bar is lowered the mower is again put in gear.

241. Draft connections.—The hitch on mowers is usually made low and below the tongue. A direct connection is sometimes made to the drag bar with a draft rod. This is styled a draw cut, and may have some advantage in applying the power more directly to the point where it is used.

242. Troubles with mowers.—If a mower fails to cut the grass and leave a clean stubble, there may be several things wrong: (1) the knife or sickle may be dull; (2) it may not fit well over the ledger plates, losing the advantages of a shear cut; (3) the knife may not register, or, in other words, it travels too far in one direction and not far enough in the other. The first of these troubles may be remedied by grinding, the second by adjusting the clips on top of the knife. There should be but a very slight clearance under these clips, and the exact amount has been given as 1/100 inch. To make the knife register in some makes, the pitman must be adjusted, while in others the yoke must be adjusted. If the mower leaves a narrow strip of grass uncut, it indicates that one of the guards has been bent down, a common thing to happen to mowers used in stony fields. Mower guards are now universally made of malleable iron and may be hammered into line with a few sharp blows with a hammer. The guards may be lined up by raising the cutter bar and sighting over the ledger plates and along the points of the guards.

243. A windrowing attachment consists in a set of curved fingers attached to the rear of the cutter bar, which rolls the swath into a windrow. It is useful in cutting clover, peas, and buckwheat. The attachment may be used as a buncher with the addition of fingers to hold the swath until tripped.

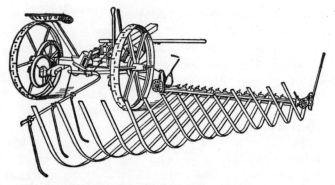

FIG. 123—A WINDROWING ATTACHMENT FOR A MOWER. IT MAY ALSO BE
USED AS A BUNCHER

244. Knife grinder.—The knife grinder is a handy tool
which may be attached to a mower wheel or to a bench.

FIG. 124—A SICKLE OR MOWER KNIFE GRINDER

It is used for sharpening the mower knives. Usually it has a double-beveled emery wheel which will grind two sections of the knife at the same time. The emery wheel is given a high rotative speed by means of gearing or sprocket wheels and chain (Fig. 124).

RAKES

245. Development.—The introduction of the mower created a demand for something better and with a greater capacity than the ordinary hand rake. As long as hand methods prevailed in the cutting of the grasses there was little need for anything better than the hand rake. The first horse rake was revolving. It did very satisfactory work when carefully handled. But later in the steel tooth rake there was found a much better tool. To Walter A. Wood Company, of Hoosick Falls, New York, is given the credit for bringing out the first spring-tooth rake. Differing from the modern tool, it was made almost entirely of wood except the teeth. The early rakes were dumped entirely by hand, but later an internal ratchet was provided on the wheels, which engaged a latch operated by the foot, and which carried the rake teeth up and over, thus dumping the load. The early rakes were almost universally provided with thills.

FIG. 125—A STEEL SELF-DUMP RAKE FOR TWO HORSES. THE TONGUE MAY BE SEPARATED INTO THILLS FOR ONE HORSE. THE TEETH HAVE ONE COIL AND CHISEL POINTS

Finally arrangements were made whereby the thills could be brought together and a tongue made for the use of a team instead of one horse.

246. The steel dump rake or sulky rake.—Although the first rakes were made of wood, there are now upon the market rakes made almost entirely of steel. The rake head to which the teeth are fastened is usually made of a heavy channel bar with a minimum of holes punched through it so as not to impair its strength.

In the selection of a rake considerable variance is offered in the choice of teeth, which may be constructed of 7/16-inch or 1/2-inch round steel, may have one or two coils at the top, be spaced $3\frac{1}{2}$ inches to 5 inches apart, and have either pencil or flat points. The choice depends somewhat upon the kind of hay to be raked.

The rake is always provided with a set of cleaner teeth to prevent the hay from being carried up with the teeth when the rake is dumped. The outside teeth are sometimes provided with a projection which prevents the hay from being rolled into a rope and scattered out at the ends when the hay is very light. Sometimes an extra pair of short teeth is provided to prevent this rolling.

247. Self-dump rakes are always provided with a lever for hand dumping. Rakes are made from 8 to 12 feet in width. In the purchase of a rake the important things to look for are ease in operation, strength of rake head and wheels. Often the wheels are the first to give way. Some wheels are very bad about causing the hay to wrap about the hub. The wheel boxes should be interchangeable so they may be replaced when worn.

248. Side-delivery rakes.—The side-delivery rake was brought about by the introduction of the hay loader, the loader creating a demand for a machine which would place the hay in a light windrow. The first of these machines was manufactured by Chambers, Bering, Quinlan Company, of Decatur, Illinois.

249. One-way rakes.—Practically all of these machines consist of a cylinder mounted obliquely to the front. They carry flexible steel-wire fingers, which revolve under and to the front. These fingers roll the hay ahead, and also

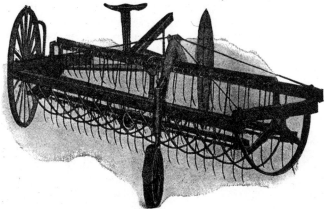

FIG. 126—ONE-WAY SIDE-DELIVERY RAKE

to one side. Some variance is to be found in the methods employed to drive the cylinder. Both gears and chain-and-sprocket drives are used.

250. Endless apron, reversible rakes.—There are other machines upon the market with a carrier or endless apron upon which the hay is elevated by a revolving cylinder and carried to either side. This machine does very satisfactory work and will place in one windrow as many as six swaths of the mower. By manipulation of the clutch driving the apron, this machine may be made to deposit the hay in bunches to be placed in hay cocks or loaded to a wagon by a fork.

The side-delivery rake takes the place of the hay tedder

to a large extent. The method of curing hay, especially clover, by raking into light windrows shortly after being mown, has proved very successful. A first-class quality of hay is obtained and in an equal length of time. It is claimed that if the leaves are prevented from drying up, they will aid very greatly in carrying off the moisture from the stems. Green clover contains about 85 per cent of water. When cured, only about 25 per cent is left. The leaves draw this moisture from the stems, and if free circulation of air is obtained the hay will dry quicker than if this outlet of the moisture for the water was cut

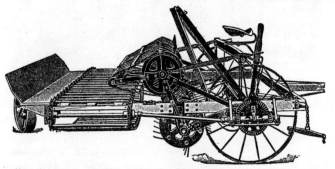

FIG. 127—THE ENDLESS APRON OR REVERSIBLE SIDE-DELIVERY RAKE

off by letting the leaves dry up. Many of the one-way side-delivery rakes may be converted into tedders by reversing the forks and the direction of their movement. The standard width for side-delivery rakes is eight feet. They are drawn by two horses.

HAY TEDDERS

251. Hay tedders.—Where a heavy swath of hay is obtained, some difficulty is experienced in getting the hay thoroughly cured without stirring. To do this stirring the hay tedder has been devised. Grasses, when cut with a mower, are deposited very smoothly, and the swath is packed somewhat to the stubble by the passing of the team and mower over it. The office of the tedder

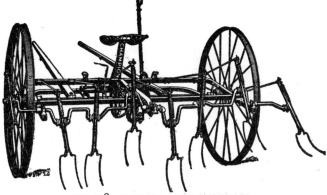

FIG. 128—AN EIGHT-FORK HAY TEDDER

is to reverse the surface and to leave the swath in such a loose condition that the air may have free access and thus aid in the curing.

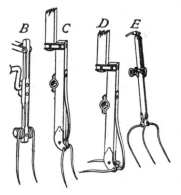

FIG. 129—TYPES OF TEDDER FORKS WITH COIL AND FLAT RELIEF SPRINGS. *D* SHOWS THE SPRING OF *C* SPRUNG

The hay tedder consists of a number of arms with wire tines or fingers at the lower ends. These are fastened to a revolving crank near the middle and to a lever at the other end. The motion of the cranks causes the tines to kick backward under the machine, thus engaging the mown hay, tossing it up and leaving it in a very loose condition. The modern machine, made

almost entirely of steel, is illustrated in Fig. 128. The
size of tedders is rated by the number of forks. Tedders
constructed of wood are still upon the market. The fork
shaft may be driven by a chain or by gearing.

HAY LOADER

252. **Development.**—The hay loader has been upon the
market for some time, but only during recent years has
there been any great demand for the tool. The Keystone
Manufacturing Company, of Sterling, Illinois, began ex-

perimenting with the hay loader as early as 1875. The
machine is designed to be attached to the rear of the
wagon, to gather the hay and elevate it to a rack on the
wagon.

253. Fork loader.—In all of the early machines the hay was placed upon the elevating apron by tines or forks attached to oscillating bars extending up over the load. The hay was pushed along this apron by these oscillating bars with the tines on the under side. This form of loader worked very satisfactorily, but had one disadvantage in working in clover and alfalfa. The oscillating bars were unsatisfactory, as they shook the leaves out of the hay. This led to the introduction of an endless apron, which works very satisfactorily in this respect. The loader equipped with oscillating forks is of much more simple construction than the other type. It also has an advantage in being able to draw the swath of hay together at the top, and force it upon the wagon. Loaders of this kind are made without gears by increasing the throw of the forks. These machines have not as yet demonstrated their advantages.

254. Endless apron loaders.—The hay is elevated in this type of loader on an endless apron or carrier after it has been gathered by a gathering cylinder. The main advantage of this type of loader is that it does not handle the hay as roughly as the fork loaders. This is an important feature in handling alfalfa and clover, as there is a tendency to shake out many of the leaves, a valuable part of the hay. Due provision must be made, however, to prevent the hay from being carried back by the carrier returning on the under side. The apron or carrier usually passes over a cylinder at the under side, which has teeth to aid in starting the hay up the carrier.

Provision must be made to enable the gathering cylinder to pass over obstructions and uneven ground. For this reason the gathering cylinder is mounted upon a separate frame and the whole held to the ground by suitable springs. The loader has a great range of capacity.

All modern machines will load hay from the swath or the windrow, and the carrier will elevate large bunches of hay without any difficulty.

FIG. 131—AN ENDLESS APRON OR CARRIER HAY LOADER

MACHINES FOR FIELD STACKING

255. Sweep rakes.—Where a large amount of hay is to be stacked in a short time, the sweep rake and the hay stacker will do the work more quickly than is possible by any other means. The sweep rake has straight wooden teeth to take the hay either from the swath or windrow, and is either drawn between the two horses or pushed ahead. When a load is secured the teeth are raised, the load hauled and placed upon the teeth of the stacker and the rake backed away.

There are three general types of sweep rakes: (1) the wheelless, with the horses spread to each end of the

FIG. 132—A TWO-WHEEL SWEEP RAKE. THE TEETH ARE RAISED BY THE DRIVER SHIFTING HIS SEAT

rake; (2) the wheeled rake, with the horses spread in the same manner; and (3) the three-wheel rake, with the horses directly behind the rake and working on a tongue.

FIG. 133—A THREE-WHEEL SWEEP RAKE. THE DRIVER IS AIDED IN LIFTING THE LOADED TEETH BY THE PULL OF THE HORSES

The latter are the more expensive. They offer advantages in driving the team, but are a little difficult to guide (Figs. 132 and 133).

256. Hay stackers are made in two general types: the **overshot** and the **swinging** stacker. In the overshot the

FIG. 134—A PLAIN OVERSHOT HAY STACKER

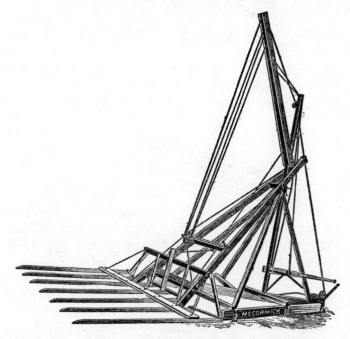

FIG. 135—THE SWING HAY STACKER. NOTE THE BRAKE AT THE REAR
END FOR HOLDING THE ROPE

teeth carrying the load are drawn up and over and the load is thrown directly back upon the stack, the work being done with a horse or a team of horses by means of ropes and suitable pulleys (Fig. 134).

The swinging stacker permits the load to be locked in place after it has been raised from the ground to any height and swung to one side over the stack. When over the stack, the load may be dumped and the fork swung back and lowered into place. The latter stackers are very handy, as they may be used to load on to a wagon. They have not as yet been built strong enough to stand hard service.

257. Forks.—A cable outfit may be arranged with a carrier and fork for field stacking, the cable being stretched between poles and supported with guy ropes. This outfit works the same as the barn tools to be described later. Very high stacks may be built by this method.

A single inclined pole may be used in stacking by raising the fork load to the top and swinging over the stack. This is usually a home-made outfit, with the exception of fork and the pulleys.

BARN TOOLS

258. Development.—The introduction of the field haying tools created a demand for machinery for the unloading of the load of hay at the barn, and this led to the development of a line of carriers and forks, the first of which was a harpoon fork, a patent for which was issued to E. L. Walfer, September, 1864. In 1873 a Mr. Nellis patented a locking device, which has given to this fork the name of Nellis fork.

J. E. Porter began the manufacture of a line of carriers

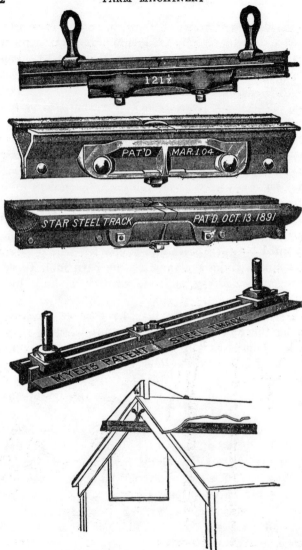

FIG. 136—TYPES OF STEEL AND WOOD HAY CARRIER TRACKS

and hay tools at Ottawa, Illinois, in 1868. This firm is still doing business. P. A. Meyers was another pioneer in the hay tool business, and in 1866 patented a double track made of two T-bars. In 1887, J. E. Porter placed upon the market a solid steel rail.

259. Tracks.—A large variety of tracks is to be found upon the market to-day—the square wooden track, the two-piece wooden track, the single-piece inverted T steel track, the double steel track made of two angle bars, and various forms of single- and double-flange steel tracks. Wire cables are used in outdoor work.

Various forms of track switches and folding tracks are to be found upon the market. By means of a switch it is possible to unload hay at one point and send it out in four different directions. In circular barns it is possible to arrange pulleys in such a way that the carrier will be carried around a circular track.

260. Forks are built in a variety of shapes and are known as single-harpoon or shear fork, double-harpoon fork, derrick forks, and four-, six-, and eight-tined grapple forks. To replace the fork for rapid unloading of hay,

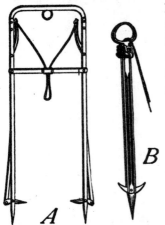

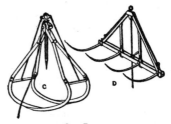

FIG. 137—*A*, DOUBLE-HARPOON HAY FORK. *B*, SINGLE-HAR-POON HAY FORK

FIG. 138—*C*, A FOUR-TINED GRAPPLE FORK. *D*, A DER-RICK FORK

the hay sling is used. The harpoon forks are best adapted for the handling of long hay, like timothy. For handling clover, alfalfa, and the shorter grasses, the grapple and derrick forks are generally used. The derrick fork is a popular style for field stacking in some localities. Harpoon forks have fingers which hold the hay upon the tines until tripped. The tines are made in lengths varying from 25 to 35 inches, to suit the conditions. The grapple fork opens and closes on the hay like ice tongs. The eight-tined fork is suitable for handling manure.

The hay sling consists of a pair of ropes spread with wooden bars and provided with a catch, by which it may

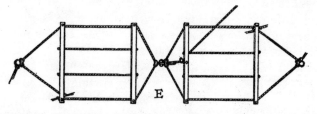

E

FIG. 139—A HAY SLING. THE SPRING CATCH BY WHICH THE SLING IS PARTED IS ABOVE E

be separated at the middle for discharging a sling load. The sling is placed at the bottom of the load, and after sufficient hay has been built over it for a sling load, another sling is spread between the ends of the hay rack and another sling load is built on, and so on. Four slings are usually required for an ordinary load; however, the number has been reduced to three, and even two. The sling is a rapid device, but is somewhat inconvenient in the adjusting of the ropes and placing in the load. It is very convenient at the finish, as the load is cleaned up well from the wagon rack, requiring little hand labor. The most popular method at the present time is to use forks to remove all the load but one slingful, which is removed by a sling placed in the bottom of the load. This method circumvents the necessity of building slings into the load or hand labor in cleaning up the load for the fork at the finish. If the standard sling carrier is used, it is necessary

FIG. 140—A TWO-WAY FORK HAY CARRIER. TO WORK IN THE OPPOSITE
DIRECTION, THE ROPE IS SIMPLY PULLED THROUGH UNTIL
THE KNOT ON THE OPPOSITE END IS STOPPED
BY THE CARRIER

to use two forks; however, a special fork and sling carrier will permit the use of a single fork.

261. Carriers.—Carriers are made to suit all of the various forms of tracks and are made one-way, swivel,

and reversible. In order to work the one-way from both
ends of a barn it is necessary to take it off the track and
reverse. The swivel needs only to have the rope turn to
the opposite direction, while in the reversible the rope
is knotted at each end, and when it is desired to work

FIG. 141—A DOUBLE-CARRIAGE REVERSIBLE SLING CARRIER. DESIGNED FOR
HEAVY SERVICE

from the other end of the barn all that is necessary is
simply to pull the rope through the other way.

There are numerous devices to be used with
barn outfits, carrier returns, pulley-changing devices,
which are very handy, but need only be mentioned
here.

BALING PRESSES

262. Development.—Many patents were granted on baling presses during the early half of the past century, indicating the rise of the problem of compressing hay into a form in which it could be handled with greater facility. It was not, however, until 1853 that H. L. Emery, of Albany, N. Y., began the manu-

FIG. 142—A LIGHTER SLING CARRIER LOADED WITH A SLING LOAD OF HAY

facture of hay presses. It is stated that this early machine had a capacity of five 250-pound bales an hour and required two men and a horse to operate it. It made a bale 24 × 24 × 48 inches.

The next man to devote his efforts toward the development of a hay press with any success was P. K. Dederick, who began his work about 1860. He produced a practical hay press.

George Ertel was the pioneer manufacturer of hay presses in
the West. His first efforts were in 1866, and from that time he
devoted practically his entire time to the manufacture of hay
presses. His first machine was a vertical one operated by horse
power. Now both steam and gasoline engines are used to
furnish the power.

263. Box presses are used very little at present, being
superseded by the continuous machines of larger capacity.
The box press consists in a box through which the
plunger or compressor acts vertically, power being furnished either by hand or by a horse. The box, with the
plunger down, is filled with hay; the plunger is then
raised, compressing the hay into, usually, the upper end,
where it is tied and removed. The machine is then prepared for another charge.

264. Horse-power presses are either one-half circle or
full circle. In the half-circle or reversible-lever presses

FIG. 143—A FULL-CIRCLE HORSE HAY PRESS ON TRUCKS FOR
TRANSPORTATION

the team pulls the lever to one side and then turns around
and pulls it to the other side. The hay is placed loose
in a compressing box, compressed at each stroke and
pushed toward the open end of the frame, where it is
held by tension or pressure on the sides. When a bale
of sufficient length is made a dividing block is inserted
and the bale tied with wire.

In the full-circle press the team is required to travel
in a circle. Usually two strokes are made to one round

of the team. Various devices or mechanisms are used to
obtain power for the compression. It is desired that the
motion be fast at the beginning of the stroke, while the
hay is loose, and slow while the hay is compressed during
the latter part of the stroke. The cam is the most common device to secure this; however, gear wheels with a

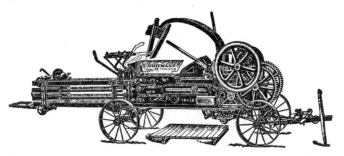

FIG. 144—A HAY PRESS FOR ENGINE POWER AND EQUIPPED WITH A CON-
DENSER TO THRUST THE HAY INTO THE HOPPER

cam shape are often used. The rebound aided by a spring
is usually depended upon to return the plunger for a new
stroke; but a cam motion may be made use of to return
the plunger. It is to be noted that some machines use a
stiff pitman and push away from the power, while others
use a chain and rod and pull the pitman toward the power
or reverse the direction of travel of the plunger. A horse-
power machine has an average capacity of about 18 tons
a day. A cubic foot of hay before baling weighs 4 or 5
pounds when stored in the mow or stack. A baling press
increases its density to 16 or 30 pounds a cubic foot.
Specially designed presses for compressing hay for export
secure as high as 40 pounds of hay a cubic foot.

265. Power presses make use of several variable-speed
devices and a flywheel to store energy for compression.
Power machines are often provided with a condenser to

thrust the hay into the hopper between strokes. The common sizes of bales made are 14 × 18, 16 × 18, and 17 × 22 inches in cross-section, and of any length. A new baler has appeared which is very rapid, making round bales tied with twine. The machine can readily handle the straw as it comes from a large thresher. Plunger presses are built with a capacity up to 90 tons a day.

CHAPTER IX

MANURE SPREADERS

266. Manure as a fertilizer.—Although the manure spreader has been a practical machine for some time, it is only recently that its use has become general. This is especially true in the Middle West, where for a long time the farmer did not realize the need of applying manure, owing to the stored fertility in the soil when the native sod was broken, and cultivated crops grown for the first time. It has been proved that manure has many advantages over commercial fertilizer for restoring productiveness to the land after cropping. It has been estimated by experts of the United States Department of Agriculture that the value of the fertilizing constituents of the manure produced annually by a horse is $27, by each head of cattle $19, by each hog $12. The value of the manure a ton was also estimated at $2 to $7. It is not known from what data these estimates were made. The value of manure as a fertilizer does not depend solely upon the fact that it adds plant food to the soil, but its action renders many of the materials in the soil available and improves the physical condition of the soil.

267. Utility of the manure spreader.—As it was with the introduction of all other machines which have displaced hand methods, there is much discussion for and against the use of the manure spreader. The greatest advantage in the use of the manure spreader lies in its ability to distribute the manure economically. Experiment has shown that, in some cases at least, as good results can be obtained from eight loads of manure to the acre as twice that number. It is impossible to distribute and spread by hand in as light a distribution as by the spreader. The manure is thoroughly pulverized and not spread in large bunches, which become fire-fanged and of little value as a fertilizer. It is a conservative statement that the manure spreader will make a given amount of manure cover twice the ground which may be covered with hand spreading. Since a light distribution may be secured, it can be applied as a top dressing to growing crops, such as hay and pasture, without smothering the crop. The manure spreader also saves labor. It is capable of doing the work of five men in spreading manure. With a manure loader or a power fork it is possible to handle a large amount of manure in a short time.

268. Development.—The first attempts at the development of a machine for automatically spreading fertilizer were contemporaneous with a machine for planting or seeding. In 1830 two brothers, by the name of Krause, of Pennsylvania, patented a machine for distributing plaster or other dry fertilizer. This machine consisted of a cart with a bottom sloping to the rear, where a transverse opening was provided with a roller underneath. This roller was driven by a belt passed around one of the wheel hubs. It fed the fertilizer through the opening.

The first apron machine was invented by J. K. Holland, of North Carolina, in 1850. The endless apron was attached to a rear end board and passed over a bed of rollers and around a shaft driven by suitable gearing at the front end of the cart. After the box had been filled with fertilizer and the apron put in gear, it drew the fertilizer to the front and caused it to drop little by little over the front end.

The first spreader of the wagon type was produced by J. H. Stevens, of New York, in 1865. His machine had an apron which was driven rearward by suitable gearing to discharge the load and was cranked back into position for a new load. The later machines were provided with vibrating forks at the rear end.

which fed the manure to fingers extending to each side, and securing in this way a better distribution of the fertilizer than the former ways. Thomas McDonald, in 1876, secured a patent on a machine much like the Stevens machine, except that it was

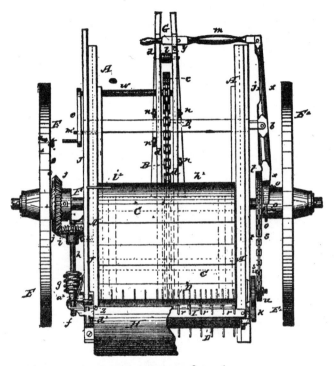

FIG. 145—THE J. S. KEMP MACHINE OF 1877. (FROM A PATENT OFFICE DRAWING)

provided with an endless apron passing around the roller at each end of the vehicle.

Many of the ideas of the modern spreader made their appearance in the patent of J. S. Kemp, granted in 1877. The objects of the invention read as follows: "To provide a farm wagon or cart with a movable floor composed of slats secured to an endless belt or chain. To the foremost slat an end board is secured, which, when the machine is in forward motion, moves by a suitable gearing slowly to the rear, thus propelling the material that may be loaded in the vehicle against a rotating toothed drum, which pulverizes and evenly spreads the load on the ground behind."

A spreader with a solid bottom to the box over which the manure was drawn by chains with slats across and attached to an end board, appeared in 1884. Variable-speed devices for varying the rate of distribution were provided at the same time.

An endless apron machine appeared in 1900, with hinged slats which overlapped while traveling rearward, and which hung downward while traveling ahead on the under side, making an open apron. There is a tendency on the part of endless apron machines to become fouled by the manure which passes through the apron on the upper side and lodges on the inside of the lower half.

It would be impracticable to mention all of the improvements to manure spreaders along the line of return motions, variable-feed devices, safety end boards, and almost countless details in the construction of bed, apron, and beater.

THE MODERN SPREADER

The modern manure spreader consists essentially in (a) a box with flexible apron for a bottom, (b) gearing to move the apron to the rear at a variable speed, and (c) a toothed drum or beater to pulverize and spread the manure evenly behind.

269. Aprons.—Three types of aprons or box bottoms are to be found in use on the modern spreader: (a) a **return apron** (Fig. 146), with an end board which pulls the load to the beater by being drawn under the box; (b) the **endless apron** (Fig. 147), which is composed of slats or bats passing continuously around reels at each end of the box; and (c) bars or a push board, moved by chains, thus moving the load to the beater over a solid floor.

The endless apron spreader is perhaps of more simple construction than the others, as no return motion is needed to return the apron for another load. It will not distribute the load well at the finish because it does not

FIG. 146—A RETURN APRON SPREADER, SHOWING THE APRON UNDER-
NEATH, AND ALSO A GEAR AND CHAIN DRIVE TO BEATER

have the end board to push the last of the load to the beater. There is also some difficulty in preventing the inside of the apron from being fouled with manure. One make overcomes this difficulty by hinging the slats in a

FIG. 147—AN ENDLESS APRON

way that they may hang vertically while on the lower side. To prevent fouling, the endless apron may be covered with slats for only half its length. The chain apron without doubt requires much more power than the others, since the weight is not carried upon rollers. Some spreaders have an advantage over others in the arrangement of rollers and the track on which they roll. The rollers may be either attached to the bed or to the slats.

270. Main drive.—The main drive to the beater varies with different machines. The power may be taken from the main axle with a large gear wheel or by means of a large sprocket and a heavy chain or link belt. It is

FIG. 148—A CHAIN DRIVE TO THE BEATER. NOTE THE METHOD OF REVERSING THE MOTION

almost universal practice to use a combination of a chain and a gear in the drive. The speed of the beater must be such that the speed of the traction wheels must be increased twice, while the direction of rotation must be reversed. To reverse the direction of the motion, the gear is used. The heavy chain or link belt offers some advantage in case of breakage. A single link may be replaced at a small cost, while if a tooth is broken from a large gear the entire wheel must be replaced.

The use of gears is avoided entirely in at least one make by passing the drive chain over the top of the main sprocket and back instead of around it. This reverses the direction of rotation (Fig. 148). Some spreaders are so arranged that a large part of the main drive must be kept in motion even when the machines are out of gear. The gearing must be well protected, **or it**

FIG. 149—A CHAIN AND GEAR DRIVE TO THE BEATER. THE BEATER IS PLACED IN GEAR BY MOVING BACK UNTIL GEARS MESH

will become fouled in loading. The **main axle** must be very heavy on a spreader, as a large share of the load is placed upon it, and it must not spring or it will increase the draft greatly. Large bearings should be provided with a reliable means of oiling and excluding dirt.

271. Beaters.—The beater is usually composed of eight bars filled with teeth or pegs for tearing apart and pulverizing the manure (Fig. 150). Some variance is noticed in the diameter of the beater and its location as to height. It is claimed by certain manufacturers that much power

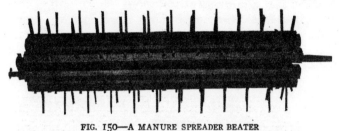

FIG. 150—A MANURE SPREADER BEATER

may be saved by building the beater large and placing it low; in this way there is no tendency to compress the manure on the lower side of the beater, as it is not necessary to carry the manure forward and up. When a

FIG. 151—A MANURE SPREADER WITH AN END BOARD TO BE PLACED IN FRONT OF THE BEATER

beater is so placed it does not have the pulverizing effect it would have otherwise. When a load is placed upon a spreader it is usually much higher and more compact in the center. If due provision is not made, the spreader will spread heavier at the center than at the sides. One beater has the teeth arranged in **diagonal rows,** tending to carry the manure from the center to the sides. Several have **leveling rakes** in front of the beater, and at least one a vibrating rake, to level and help pulverize the manure. If no provision is made, the front of the beater will be filled with manure while loading, and the

FIG. 152—A RATCHET DRIVE FOR THE APRON. NOTE METHOD OF VARYING THE FEED

machine will not only be difficult to start, but will carry over a heavy bunch of manure when put in motion. To surmount this difficulty, the beater in some makes is made to move back from the load when put in gear. A few machines have an end board, which is dropped in front of the beater while the load is put on, and lifted when spreading is begun.

272. Apron drives.—At least two systems of apron drives are in use: (*a*) the ratchet, and (*b*) the screw or

worm gear drive, the feed being regulated in the latter
case with a face gear or cone gears and a flexible shaft.
The **ratchet drive** (Fig. 152) has an advantage in offering
a great range of speed. As many as ten speeds for the
apron, or in reality ten rates of feed, may be obtained.
However, the motion is intermittent and heavy strains
are thrown upon the driving mechanism by the sudden
starting of the heavy load. The ratchet drive is liable
to breakage and does not prevent the load from feeding
too fast in ascending a hill owing to the tendency of the

FIG. 153—THE WORM GEAR DRIVE TO THE APRON.   ALSO FACE GEAR FOR
VARYING THE FEED

load to run back. To prevent this a brake is used, but
must be unsatisfactory.

The worm drive, on the other hand, gives a constant
motion to the apron, but does not offer a great variety of
feeds, and unless carefully attended to wears out quickly.
Fig. 153 shows a worm drive with a face gear for varying the feed. The worm drive must be greased several
times each day or it will cut out. It has been known for
a worm gear to wear out in a single day's work. The
cone gear for varying the speed is very little used, but
seems to be a satisfactory drive.

The return motion is usually independent of the forward motion, a safety device being arranged to prevent both forward and return motions being put in gear at the same time. In the early machines the apron was returned by hand, but now power is universally used. A crank is sometimes provided by which the apron may be returned by hand if desired. The endless apron, of course, requires no return motion.

273. Wheels.—At the present time there is some discussion in regard to the merits of wood and metal wheels for manure spreaders. The large cast hub needed to carry the driving pawls or the main ratchet is favorable to the use of a wood wheel. This type of wheel has been displaced on practically all other implements, and it is safe to venture an opinion that it will be displaced in time on the manure spreader. Wide tires of 5 or 6 inches are essential on the manure spreader. In order to secure greater traction the wheels must often be provided with **grouters** or **traction bands.** The traction band may be removed when not needed, permitting the spreader to travel more smoothly over hard ground.

274. Trucks.—As now constructed, the manure spreader has a low front truck arranged to turn under the bed. A low truck offers an advantage in loading, but undoubtedly is of heavier draft. A narrow front truck prevents a lashing of the neck yoke in passing over uneven ground.

275. The frame of a manure spreader must be constructed of good material, and should also be well braced and trussed with iron rods. Not only must the material be strong, but also able to resist the rotting action of the manure.

276. Simplicity.—It is desirable that the manure spreader as well as other machines shall be as simple as possible. Multiplied systems of gearing and levers are not desirable on any machine. The best results are obtained from few working parts, provided they will do the work.

277. Sizes.—The capacity of manure spreaders is given in bushels, yet there appears to be very little connection between the bushel and capacity of manure spreaders. By measuring, it has been found that certain spreaders' capacity would be more nearly correct if given in cubic feet instead of bushels.

278. Drilling attachment.—To apply manure to growing crops planted in rows and to economize the manure, a drilling attachment is provided. It consists in a hood for the beater, with funnels below, from which the manure is discharged beside or on each row. The attachment may also be used to distribute lime and other fertilizers.

279. Other uses.—The manure spreader may be used to distribute straw and other material for mulching. With the beater removed, the manure spreader may be used as a dump wagon for hauling and dumping stone, gravel, etc. It is especially useful in hauling potatoes and root crops where they are to be dumped into a chute leading to the root cellar. The apron in this case is moved back by hand power by means of a crank provided for the purpose.

CHAPTER X

THRESHING MACHINERY

280. Development.—In the oldest of writings mention is made of the crude devices by which grain in the ancient times was separated from the straw. Although mention of mechanical devices was made at a very early time, the two methods which came into extended use were treading with animals and beating the grains from the ears with a flail. The flail was nothing more nor less than a short club usually connected to a handle with a piece of leather. This long handle enabled the operator to remain in an upright position and strike the unthreshed grain upon the floor a sharp blow. After the grain was threshed from the head or ear, the straw was carefully raked away and the grain separated from the chaff by throwing it into the air and letting the wind blow out the chaff, or by fanning while pouring from a vessel in a thin stream. Later a fanning mill was invented to separate the grain from the chaff.

Flailing was the common method of threshing grain as late as 1850. In regard to the amount of grain threshed in a day with a flail, S. E. Todd makes the following statement in Thomas's book on Farm Machinery: "I have threshed a great deal of grain of all kinds with my own flail, and a fair average quantity of grain that an ordinary laborer will be able to thresh and clean in a day is 7 bushels of wheat, 18 bushels of oats, 15 bushels of barley, 8 bushels of rye, or 20 bushels of buckwheat."

281. Early Scotch and English machines.—About the year 1750 a Scotchman named Michael Menzies devised a machine which seems to have been nothing more nor less than several flails operated by water power. This machine was not practical, but in 1758 a Mr. Lechie, of Stirlingshire, England, invented a machine with arms attached to a shaft and inclosed in a case. Lechie's machine gave the idea for the more successful machines which came later.

A Mr. Atkinson, of Yorkshire, devised a machine (the date is

FIG. 154—A THRESHING MACHINE IN OPERATION. THE GRAIN IS SUPPLIED TO THE MACHINE FROM ONE SIDE IN ORDER TO OBTAIN A BETTER VIEW OF THE MACHINE

not known) having a cylinder with teeth, or a peg drum, as it was called, and these teeth ran across other rows of teeth, which acted as concaves.

282. American development.—The Pitts brothers have figured more prominently than any two other men in the early development of threshing machines in America. Others were granted patents, but to these men credit should be given for inventing and manufacturing the first practical machine. These brothers were Hiram A. and John A. Pitts, of Winthrop, Maine. A patent was granted to them December 29, 1837, on a thresher, the first of the "endless apron" type. This machine was made not only to thresh the grain, but to separate it from the straw and the chaff. Although this machine as constructed by the Pitts brothers was different from the modern separator, it contained many of the essential features. It had but a single apron. The tailings elevator returned the tailings behind the cylinder over

FIG. 155—THRESHING MACHINE OF 1867

the sieves to be recleaned, instead of into the cylinder, as now arranged.

In the Twelfth Census Report, the following statement is made: "The first noteworthy threshing or separating machine invented in the United States which was noticeable was that of Hiram A. and John A. Pitts, of Winthrop, Maine, and may be said to be the prototype of the machines used at the present time."

The first machines and horse powers to drive them were satisfactory. The machine was finally made so it could be loaded on trucks, transported from place to place, and set by removing the trucks and staking to the ground. This type of machine received the name of "groundhog thresher." Later the machines were mounted on wheels, and hence were quite portable. The early horse power consisted of a vertical shaft mounted between beams, to which a sweep was attached. The power was taken by a tumbling rod from a master wheel mounted above. This type earned the name of "cider-mill" power. Tread power was used largely to operate the early threshers, and water power to some extent. John A. Pitts finally located a factory at Buffalo, New York, and the "Buffalo Pitts" thresher became well known throughout the country. This machine is manufactured to-day with some of the original features. John Pitts died in 1859. Hiram A. Pitts moved to Chicago in 1852 and established a factory which built what was known as the "Chicago Pitts." He died in 1860. Much credit is due to these men for the development of a practical threshing machine.

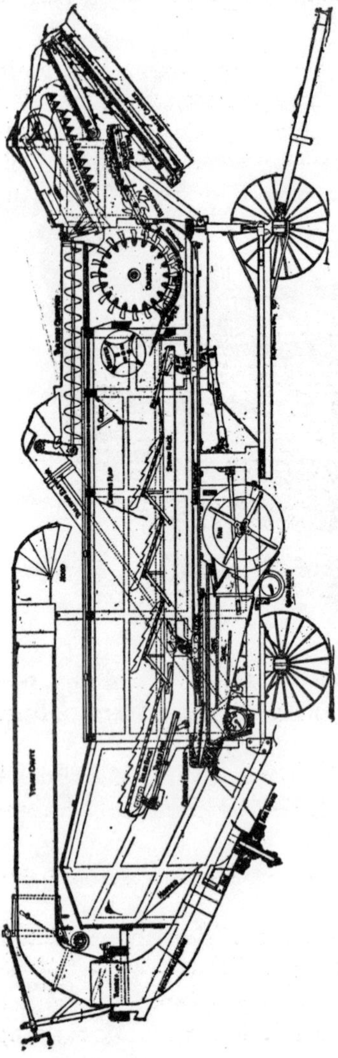

FIG. 156—SECTION OF A MODERN THRESHING MACHINE

THE MODERN THRESHING MACHINE OR SEPARATOR

283. Operations.—The threshing machine as it is constructed to-day performs four distinct operations. These operations and the parts that are called upon to perform them in most machines may be enumerated as follows:

First, shelling the grain from the head. The parts which do the shelling or the threshing are the cylinder and the concaves with their teeth. Fig. 157 shows these parts.

Second, separating the straw from the grain and chaff. The parts which perform this operation are the grate, the beater, the checkboard, and the straw rack, or raddle.

FIG. 157—THE CYLINDER AND CONCAVE

Third, separating the grain from the chaff and dirt, performed by the shoe, fan, windboard, screens, and tailings elevator.

Fourth, delivering the grain to one place and the straw to another, which is accomplished by the grain elevator and the stacker or straw carrier.

Other attachments, as the self-feeder and weigher, are often provided.

These parts will now be discussed somewhat in detail.

284. Cylinder.—The cylinder is usually made by attaching parallel **bars** to the outer edge of **spiders** mounted on the **cylinder shaft.** The whole is made very rigid by shrinking wrought-iron bands over the bars. A solid cylinder may be used instead of the bars. The bars in some makes are made of two pieces, and hence are called **double-barred** cylinders. The teeth are held in place by nuts or wedges, and are often provided with lock washers. Wooden bars may be placed under the nuts and act as a cushion, preventing the teeth from loosening as readily as otherwise. The cylinder has usually 9, 12, or 20 bars, the latter being spoken of as a big cylinder.

The cylinder travels with a peripheral speed of about 6,000 feet a minute. The usual speed for the 12-bar cylinder is 1,100 revolutions per minute, and of the 20-bar cylinder is 800 revolutions per minute. A large amount of power is stored in the cylinder when in motion and enables the machine to maintain its speed when an undue amount of straw enters the cylinder at a time.

The kernels of grain should be removed from the heads and retaining hulls in passing through the cylinder. The other devices in the machine do not have a threshing effect. In threshing damp, tough grain, a higher speed must be maintained than when threshing dry grain. It is attempted, however, to run the cylinder at about uniform speed in nearly all cases.

As the cylinder is heavy and travels at a high speed, it must be properly balanced, or it will not run smoothly. In the factory the cylinder is made up and then balanced by running at a high speed on loose boxes. The heavy side is located by holding a piece of chalk against the cylinder while in motion. When the cylinder teeth become worn they must be replaced with new teeth, which are heavier, so that there is a tendency to put the cylinder out of balance. After putting in new teeth the cylinder may be balanced by removing from the machine and mounting it on two level straight edges placed on saw horses or trestles. Two steel carpenter's squares will answer for straight edges. The heavy side of the cylinder will be found, because it will come to rest at the lower side. Weights may be added in the shape of nuts and wedges to bring the cylinder into balance. This latter method will not bring the cylinder into perfect balance, as one end may be heavy on one side, while at the opposite end the other side will be the heavier, and the cylinder will appear to be in perfect balance on the straight edges.

The cylinder must have end adjustment in order that its teeth will travel directly between the concave teeth. If the cylinder teeth travel close to the concave teeth on one side they will crack the kernels and break up the straw, and thus leave a larger opening on the opposite side through which the grain may pass unthreshed. It is advisable that the cylinder shaft be heavy and equipped with self-aligning boxes provided with a reliable oiling device. Some machines are made with an "outboard" bearing on the pulley end of shaft, i. e., outside of main drive pulley. This arrangement is strong but somewhat difficult to line up, and the belt cannot be detached readily.

285. The concave received its name from its shape being hollowed out to conform to the shape of the cylinder. The concave carries teeth which resemble the cylinder teeth very much, and have openings through which some of the threshed grain may fall. It is made in sections, so the number of teeth may be varied by substituting different sections. It may be moved or adjusted to or from the cylinder. In some machines the adjustment may be made at the front and at the rear independently of each other, it being claimed that an advantage is gained by having the concave lower at the rear in order that a larger opening be provided for the straw to pass through as it is expanded in the operation of threshing. As a rule, it is advisable to use few rows of concave teeth and set them well up against the cylinder, as there is little chance of the concave becoming clogged.

286. Cylinder and concave teeth.—The teeth in both the cylinder and the concave are curved backward slightly to prevent the straw being carried past the cylinder without being threshed. Teeth become more rounded by use and reduce the capacity and interfere with the proper working of the machine. It is stated that a very large amount of power is required when the teeth become rounded off. When worn the teeth should be replaced, making it necessary to balance the cylinder before replacing in the machine, and also calling for watchfulness on the part of the thresher lest some of the new teeth become loose and cause damage. The teeth are usually made of a good grade of mild steel, yet certain manufacturers prefer tool steel with a hardened edge. No doubt the latter wear better.

FIG. 158—THE GRATE AND CONCAVE

287. The grate consists of parallel bars, with openings between, designed to retard the straw and allow a large portion of the grain to pass through to the grain conveyor before reaching the straw rack (Fig. 158)

288. The beater.—After passing through the cylinder and concave and over the grate the grain comes in contact with the beater. The beater (Fig. 159) is a fan-like device which tends to carry the straw away from the cylinder and forms a stream of straw to pass over the straw rack. Some makers make use of two beaters, one above the straw and one below, in an effort to separate the grain and chaff from the straw. The beater must

FIG. 159—THE BEATER

run at high enough speed to enable the centrifugal force to prevent the straw from wrapping around it.

289. The checkboard.—The purpose of the checkboard is to stop the kernels which may be thrown from the beater. It is usually constructed of sheet iron and allowed to drag over the stream of straw.

290. Straw rack.—The straw rack is for the purpose of carrying the straw away from the cylinder and shaking

FIG. 160—ONE TYPE OF STRAW RACK

the grain down on the grain conveyor below. Straw racks are of three types: (*a*) endless apron or raddle type, (*b*) oscillating racks, (*c*) vibrating racks.

The endless apron or raddle rack consists of a web with thumpers underneath to shake the grain to the bottom. It is usually made in sections with an opening between which permits the grain and chaff to fall through. The endless apron was the first device used and is now found in only a few machines, and there only in short lengths.

The oscillating rack is made in sections and attached to a crank shaft directly. The sections are made to balance each other and offer a great advantage in this respect. An oscillator is a very good device for separating the grain, but perhaps somewhat difficult to keep in repair.

The vibrating rack may be made in one or more sections. When made in one section there is usually an attempt to balance its motion with that of the grain pan. The rack is provided with notched fingers, called "fishbacks." These are given a backward and upward thrust by a pitman attached to a crank, causing the rack to swing on its supports. This motion causes the straw to move backward and at the same time be thoroughly agitated. Machines are constructed with two racks, the upper to carry off the coarse straw and a lower to separate the finer. The double rack permits of their motion being balanced the same as the rack built in two sections.

291. The grain conveyor or grain pan extends from under the cylinder back almost the full length of the machine. Its function is to convey the grain to the cleaning mechanism. It should be of light, yet strong, construction. It must not sag, or grain will be pocketed in such a manner that its motion will not cause it to pass on.

292. Chaffer.—At the end of the grain conveyor and really forming a part of it is the chaffer, which is a sieve with large openings permitting all but the coarse straw to pass through. A part of the blast from the fan passes through the chaffer, and a large portion is carried off in this manner. At the back of the chaffer is placed the **tailings auger,** which catches the part heads and grains with the outer hulls, to return them by way of the tailings elevator to the cylinder to be rethreshed. Over the tailings auger an adjustable conveyor extension is usually placed to aid in stopping the unthreshed heads.

293. The shoe.—The shoe is the box in which the sieves are mounted, and which has a tight, sloping bottom to carry the threshed grain to the **grain auger.** The shoe is always given a motion to shake the grain through

FIG. 161—FAN, SHOE, AND CHAFFER

it. If this motion be lengthwise with the machine, it is said to have **end shake;** if across the machine, it is said to have **cross shake.** The latter is used very little at present.

294. The sieves.—The sieves consist of a wooden frame covered with woven wire cloth or a perforated sheet of metal. **Adjustable sieves** are constructed in which the size of openings may be adjusted to suit the work done. The openings in the sieve should be large enough to permit the passage of the kernel downward, and of sufficient number to permit the blast to pass upward through it. The sieve must be well enough supported so it will not sag when loaded, or the grain will settle to the low spot and clog the sieve. The frame should be strong, and perhaps reënforced with a malleable casting at each of the corners.

295. **The fan** consists of a series of blades or wings mounted on a shaft. A blast is thus created to blow the chaff from the grain. An **overblast** fan delivers the blast backward from the blades at the upper portion of the fan drum. The **underblast** fan rotates in the opposite direction and delivers the blast from the lower blades. Since there is a tendency to create a stronger blast from the center of the fan than from any other part, bands are placed in the fan by some manufacturers to distribute the blast more evenly across the width of the shoe.

ATTACHMENTS

296. **The self-feeder and band cutter.**—The work of the self-feeder is to cut the bands of the bound grain, distribute it across the mouth of the separator, and deliver it to the cylinder. To carry the bundles to the band cutters, the feeder must be provided with a carrier. A variety of carriers is found in use ranging from a solid canvas or rubber belt to two belts or link belts carrying slats. Both seem to be very satisfactory.

The band cutters may be knives attached to a rotating shaft, or knives similar to those in use upon mowers, the latter style of knife giving a chopping-like motion into the bundle, tending to draw them into the machine. It is claimed that this type is much better in remaining sharp for a longer time. It is not, however, of as simple construction.

Just before the grain enters the cylinder it is spread and more evenly distributed by the **retarders,** which also, as their name implies, prevent the grain from being drawn into the cylinder in bunches.

297. Stackers.—The straw carrier was for a long time the only means of carrying the straw away from the machine. This consisted in a chute, over the bottom of which the straw was drawn with a web. This developed from a carrier extending directly to the rear to an **inde-**

FIG. 162—A SECTIONAL VIEW OF A SELF-FEEDER

pendent swinging stacker and the **attached swinging stacker.** The former has gone out of use entirely, but the attached swinging stacker is used to some extent. It has some advantages over the wind stacker for barn work.

298. The wind stacker or blower has displaced the straw carrier to a large extent because it requires a smaller crew to operate. The wind stacker is made in many types. The fan drum is placed horizontal, inclined, or vertical; the straw may enter the fan direct or into the

blast after it has left the fan. The bevel gears by which
the fan is often driven are a source of trouble if the gears
do not mesh correctly from the beginning. They have
been known to wear out completely in a few days' work.
In order to obviate this trouble, the stacker drive belt is
often required to make the turn over two pulleys and
drive the fan direct. This method also gives some
trouble.

The wind stacker without doubt requires more power
than a straw carrier, but saves labor. It is impossible

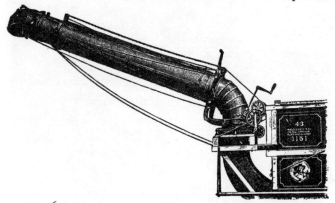

FIG. 163—A WIND STACKER. THE FAN DRUM IS NOT SHOWN

to save the straw as well, but often the straw is considered to be of little value.

299. The weigher is an attachment by which the threshed
grain is weighed and measured as threshed. It is a very
satisfactory arrangement to have on a machine doing
custom work. The weigher is nearly always provided
with an elevator by which the grain is elevated into the
wagon box. To do the elevating, pans or buckets passing through a tube are used. A few pneumatic grain elevators have been used, but not to any extent. When it

is desired to place the grain in bags a **bagger** attachment is provided, which does not elevate the grain as high.

300. Size and capacity of threshing machines.—The size of a threshing machine is indicated by the width or

FIG. 164—A WEIGHER AND BAGGER

length of the cylinder and the width of the separator proper. The two dimensions in inches are written together. The size varies from 18 × 22 inches to 44 × 66 inches, but the 32 × 54-inch or 36 × 58-inch are the common sizes. The ratio between the width of cylinder and separator varies slightly with different makes. Steam traction engines are now generally used to furnish the power for the larger sizes, although gasoline engines are being introduced into the work. A 36 × 58-inch machine requires a 15- or 16-horse-power engine, as usually rated. For the smaller sizes, horse powers and portable gasoline engines are generally used. The amount of grain threshed a day will vary very much with the conditions of the grain. There is also a wide variance in the size of machines, but the average-sized steam-operated outfit will thresh from 500 to 1,000 bushels of wheat a day or twice that number of bushels of oats.

301. Selection.—The selection of a threshing machine depends upon many conditions, among which may be mentioned the kind and quantity of grain to be threshed, the amount of labor, the power, and the condition of the bridges in the locality. There has been a gradual increase in the size of threshing outfits for some time. These large machines have an enormous capacity and require a large force of men to run them. However, the small machine is still manufactured, and there is much argument in its favor, especially so since the introduction of portable gasoline engines of a size to operate it. Steel is made use of to a large extent in the manufacture of separators, and no doubt will prove to be a very durable material when galvanized. The threshing machine deserves good care on the part of the owner. It is an expensive machine, and much money can be saved by protecting it from the weather.

302. Bean and pea threshers differ from grain threshers in having two threshing cylinders operated at different speeds. The two cylinders are necessary owing to the fact that these crops can never be cured uniformly. When the pods are dry the seeds are readily separated from the pods, and if threshed violently the seeds will split. On the other hand, when the pods are not dry the seeds cannot be separated readily and are not inclined to split. Thus in the special bean thresher the vines and pods are fed through a cylinder run at a low speed, which threshes out the dry pods. The threshed seeds are screened out, and the remaining material passes to a cylinder run at a higher speed to have the damp and greener pods

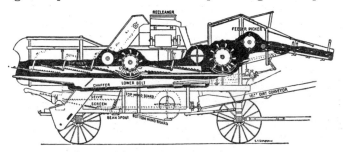

FIG. 165—SECTION OF A PEA AND BEAN THRESHER WITH TWO CYLINDERS

threshed. The bean thresher is often provided with a re-cleaner and clod crusher to remove the dirt. The size of the bean and pea threshers is indicated by the width of cylinder and the width of the separator or machine proper. Machines are usually built in the 16 × 28-, 26 × 44-, and 36 × 44-inch sizes. The larger sizes have a capacity up to 100 bushels of clean seed an hour.

303. Clover hullers resemble threshing machines very much, but differ in being provided with an additional hulling cylinder. In passing the threshing cylinder the heads are removed from the stems and the seed from the heads to some extent. The heads are separated from the stems and chaff and passed to the hulling cylinder, which removes the seed from the pods. The construction of hulling cylinders varies from a cylinder with fluted teeth and a wooden cylinder with steel brads for teeth to a cylinder covered with hardened steel rasp plates. It is necessary in all cases to have a large amount of surface

FIG. 166—SECTION OF A CLOVER HULLER

for the clover to come in contact with. Clover hullers are rated according to the size of the hulling cylinder, which may vary from 28 to 42 inches. The large machines are driven by steam power, while horse power may be used for the smaller. They may be provided with wind stackers, self-feeders, and baggers similar to threshing machines. They have a capacity up to 10 to 15 bushels of cleaned seed an hour.

CHAPTER XI

CORN MACHINERY

Feed and Silage Cutters

304. Development.—It is not an original, neither is it a novel idea, for farmers to cut dry feed for their stock. This has been going on for ages. The first machine for cutting feed was simply a knife for hacking it up. Later the feed was placed in a box, allowing the ends to come over a cutter head; then a knife was drawn down over this head, which acted in the manner of shears. Possibly the next development in feed cutters was to fasten a spiral knife to a shaft in such a manner that the cutting might be done by a continuous rotary motion. Such a cutter was invented by Mr. Salmon of England in about 1820, and by a Mr. Eastman in the United States in 1822. Another type of machine which has been developed is one in which the knives are fastened to the spokes of a flywheel, and by which the feed is chopped by being fed into the wheel, the cutting taking place over the end of the feeding board.

The storage of green and partially cured succulent crops in a silo of some form or other may be traced to the beginning of history, but it has been recently that silos have been made use of in America. In 1882 the United States Department of Agriculture could find only 99 farmers in this country who owned silos. A silo may be found on nearly every dairy farm to-day, and it is considered to be almost an essential. The silage cutter is simply the adaptation of the cutter for dry feed to the cutting of green crops.

305. Cutter heads.—Two types of cutter heads are to be found upon the market, which differ in the shape of

FIG. 167—AN ENSILAGE CUTTER WITH SELF-FEEDER AND PNEUMATIC ELEVATOR

knives used and the direction in which the fodder is fed to them. The **radial knife** is fastened directly to a flywheel,

FIG. 168—A RADIAL KNIFE CUTTER HEAD

which may also carry the fan blades for the stacker. The advantage of this type lies in the fact that it has plenty of clearance and the chopped fodder does not have any difficulty in getting away from the cutting head. The knives are usually set at an angle to give a "shear cut." To this same head short knives or teeth called splitters may be attached to split the ends of the stalk before they are cut off.

The second type of cutter head is the one which carries a **spiral knife.** The cutting edge is always the same distance from the shaft (Fig. 169). The knife may be provided with saw teeth for handling dry feed to better advantage.

306. The feeding table is provided on the larger power machines with an endless apron to carry the fodder to the feed rolls. The speed of the feed rolls and the apron is capable of adjustment for various rates of feed and coarseness of cutting.

FIG. 169—A SPIRAL KNIFE CUTTER HEAD

307. Elevators are of two general types: double-chain conveyor or **web-carrier elevator,** and the **pneumatic.** The carrier elevator is satisfactory except for very high lifts. The long webs are a source of trouble. It is economical to build silos high; hence the use of pneumatic or wind elevators. It is necessary to keep the elevator pipe almost perpendicular, or the silage will settle to one side and not be carried up by the air blast.

308. Selection.—All bearings, especially those connected with the cutting knives and feed rollers, should be very long. The shaft should be strong, and the gears heavy enough to stand a variable load. It is well to have the feed rollers so arranged that should more feed go in one side than on the other, that side could expand, yet grip the feed firmly. Since the cutter head should have a capacity of from 600 to 1,000 revolutions a minute, the frame should be made exceptionally strong and stiff. Provision should be made so the bearings cannot wind, as this causes much more friction and thus will require much more power than necessary. The capacity of silage cutters depends upon the length of each cut and upon the length of the knives, as well as the condition of the feed. In general a silage cutter should have a capacity of about one ton an hour for each horse power of power used.

HUSKERS AND SHREDDERS

309. Construction.—The husker and shredder is a combined machine to convert the coarse corn fodder, stalk and leaves, into an inviting feed for farm animals, and at the same time deliver the corn nicely husked to the bin or the wagon. By this means the entire corn crop is made use of and the fodder put into better shape for feeding.

The usual arrangement of the husker and shredder is illustrated in Fig. 170. The fodder is first placed upon

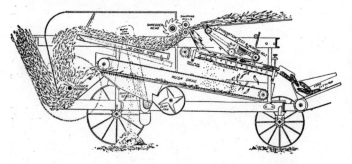

FIG. 170—SECTIONAL VIEW OF A HUSKER AND SHREDDER

the **feeding table,** from which it is fed, the butts first, to the feed or snapping rolls. Many of the machines are manufactured with self-feeders much like those for the threshing machine. Owing to the loss of hands and arms in feeding the early machines, provision is now made whereby it will be almost impossible for accidents of this nature to happen.

As the stalks pass through the snapping rolls the ears are squeezed off and allowed to fall upon a conveyor, which carries them to the husking rolls, or they may fall upon the husking rolls direct. Here the husks are pulled off and are carried to the wagon or bin. When the stalks leave the snapping rolls they pass over cutting plates and immediately are cut into small particles by the shredding head. This shredded fodder is then conveyed to the elevator, which may be either a carrier or pneumatic stacker. As the shredded fodder passes through the machine it passes over beaters, which agitate the fodder so that all shelled corn falls out and is conveyed to the wagon.

310. The snapping rolls.—The snapping rolls of the shredder may either be made corrugated, chilled, casting, or, in better machines, of tool steel, or they may be made of cast iron and with lugs inserted. The latter type seems to be well adapted to green and damp corn. The snapping rolls are given sufficient pressure by springs to grasp the stalks firmly.

311. The husking rolls rotate together in pairs, grasping the husk and tearing it away from the ears. There are very many different types of husking rolls on the market. The most common type seems to be one where the rolls are set parallel to each other in pairs. The ends of the rolls where the ear first strikes are higher than the ends where the ear leaves. Sometimes there is an apron above which forces the ears along the rolls. The devices for catching the husks are simply lugs or husking pins set in the rolls. These lugs have sharp-tempered heads. The husking rolls are held firmly together by strong springs.

312. The shredder head may be made up of several plates of steel of the rip-saw type tooth. These plates are so warped or bent that for every revolution of the head only two teeth should pass over the same point in the stock. The teeth should be offset enough to cut off a fairly good slice. In some shredders there are no **cutting plates.** The shredder head is set so close to the snapping rolls that as the stalks come through it tears them to pieces. Some machines are also provided with a revolving cutter bar.

Many machines have an interchangeable shredder and cutter head. By using the cutter head the same machine may be used in cutting straw or green fodder silage. The shredder head is also made for some machines much like a thresher cylinder, except the teeth are shorter and sharper.

313. Shelled corn separating device.—One of the essential features of a shredder is to be able to separate all shelled corn from the shredded fodder. The best means for this is to have some form of beater agitating the shredded product in the air, and thereby allowing the shelled corn to rattle through. The corn then falls through a sieve and is conveyed to a bagger or wagon elevator.

314. Size.—The size of the husker and shredder is usually denoted by the number of husking rolls, as a 4-, 8-, or 10-roll machine.

315. Capacity.—The capacity of a husker and shredder is a variable quantity, as all manufacturers will state. It is somewhat difficult to reach a definite basis upon which to rate capacity. The number of acres a day or the number of bushels a day will not state accurately the amount of work performed. In general it may be safe to state that the 8-roll husker and shredder will handle the fodder from 8 to 15 acres a day and husk from 25 to 80 bushels of corn an hour.

CORN SHELLERS

316. **Development.**—The earliest device used in the shelling of Indian corn or maize was a simple iron bar placed across a box and over which the ear of corn was rasped. The edge of a shovel was often used in place of this bar. Another early scheme was to drive the ear with a mallet through a hole just large enough to let the cob pass through.

Edmund Burke, Commissioner of Patents, in making his report for the year 1848, states that two patents were granted on corn shellers. He also states: "Corn shellers have usually been constructed in one of three modes. In the first the shelling is performed on the periphery of a cylinder; in the second it is done on the sides (one or both) of a wheel; and in the third it is done by forcing, by means of a mallet or hammer, the cob, surrounded by the corn, through a hole sufficiently large to admit the cob only. The sides of this hole are called the strippers and are often arranged in radial sectional pieces of four, six, or eight each, acting concentrically against the corn or cob by the force of a spring or substitute behind.

"To this last kind of corn sheller there have been raised several objections, the most prominent of which is that in the opening of the radial sections by stripping the corn from the cob the kernels often become entangled and wedged between the radial sections and prevent some one or more of the sectional pieces from acting upon the rows of corn to which it may be opposite."

Among the early American inventors, Clinton and Burrall are the best known. The Burrall sheller was probably most popular. It was made of iron, furnished with a flywheel to equalize velocity, and was worked by one person while another fed it. It discharged the corn at the bottom and the cob at the end. Allen Wayne was the first man to make a two-hole sheller.

317. **Types of the modern sheller.**—There are two general types of corn sheller to-day outside of the ware-

house sheller, which will not be considered here. Only portable shellers will be discussed. One will be called the **spring sheller,** and the other is the well-known **cylinder sheller.**

318. **The spring sheller.**—This term may not be generally accepted, although it is a name applied by several manufacturers to the sheller whose shelling mechanism consists in **picker wheels, bevel runners,** and **rag irons,** held in place with springs. This type of sheller is illustrated in Fig. 172. It is also called the "picker" type of sheller. The

FIG. 171—A ONE-HOLE HAND SHELLER

parts mentioned which come in contact with the corn

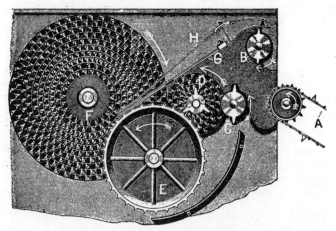

FIG. 172—SHELLING MECHANISM OF THE PICKER OR SPRING SHELLER. *A,* FEED CHAIN ; *B, C,* BEATERS ; *D, F,* PICKER WHEELS ; *E,* BEVEL RUNNER ; *G,* RAG IRON ; *H,* SPRING

are made of chilled iron and are very hard. The tension on the rag-iron springs may be adjusted and should be capable of individual adjustment when necessary. The most important advantage of the spring sheller is that it leaves a whole cob. It is especially desirable to have whole cobs where they are used for fuel.

319. The cylinder sheller.—The shelling mechanism of the cylinder sheller is shown in Fig. 173, and is described by the manufacturer as follows: "The shelling cylinder is made of heavy rods of wrought iron placed equidistant,

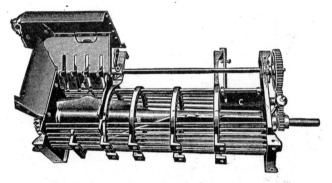

FIG. 173—SHELLING MECHANISM OF CYLINDER SHELLER

presenting a corrugated surface which cannot wear smooth. Within this a revolving iron cylinder with spiral vanes threshes the corn against the surfaces of the rod cylinder. The vanes approach the rods sufficiently close to keep every ear in rapid motion, shelling one ear or one bushel with the same facility. A regulator at the discharge end places the machine within control of the operator. The spaces between the rods allow the shelled corn to escape freely, thus lessening the draft, relieving the cylinder from clogging and from all liability to cut or grind the grain." The cylinders are made adjustable to suit various sizes of corn.

320. Self-feeder.—The purpose of the self-feeder is to carry the ears to the shelling mechanism. The spring shellers are provided with feeder chains, which carry teeth to "end up" the ears and carry them directly to each set of shelling wheels, or to each "hole," as it is

FIG. 174—A SIX-HOLE POWER SHELLER

called. The cylinder sheller uses a double chain conveyor with slats between, as it is not necessary to end up the ears. In all spring shellers provision must be made for forcing the ears into the holes. This is accomplished by adding **picker-feeding wheels** or a **beater.**

321. Extension feeders.—In shelling corn from large cribs, extension feeders are provided to circumvent the carrying of the corn by hand. These are provided with double-chain conveyors and may be had in sections, making a "drag conveyor" which may be extended to almost any direction from the main feeder.

322. Separating device.—To separate the corn and the cobs, the whole, after passing through the shelling mechanism, is made to pass over a cob rack which permits the corn and chaff to pass through. The cob rack is made in at least three ways—a vibrating rack, a rod rack with rakes, or an endless rack with thumpers underneath. The latter two have advantage in lightness and amount of power required, and also in the steadiness by which the machine may be operated.

323. Cleaning device.—To clean the corn and free it from chaff and husks a fan is provided which sends its blast through some form of sieve or rack. The corn sieve may be dispensed with and a single rack used.

324. Grain elevator.—The grain on all portable machines is elevated by a chain cup elevator into the wagon box. To carry the corn to the lower end of the elevator an auger is universally used.

325. Cob carrier.—To carry the cobs from the sheller a single- or double-chain conveyor is used. It is an advantage to have this swing from the sheller.

326. Dustless sheller.—To carry the chaff and husks away from the sheller an auxiliary fan is provided on the larger machines to gather and discharge the dust and chaff at one point. A sheller so arranged is called a dustless sheller.

327. Shuck sheller.—A few of the spring shellers are arranged to handle partially husked corn, and many of the cylinder shellers are so arranged. The capacity of the machine is much reduced in handling snapped or unhusked corn.

328. Power.—The power required for a four-hole spring sheller is usually about eight horse. The six-hole machine requires about 10 and the eight-hole 12 to 14 horse power. The power required for cylinder shellers varies with the style and manufacturer's number.

329. Capacity.—The capacity of the spring sheller is determined by its size, which is denoted by the number of holes, which vary from the one-hole hand-power machine to the large eight-hole power sheller. A four-hole sheller is usually rated at 100 to 200 bushels an hour, the six-hole at 200 to 300, and the eight-hole at 300 to 600 bushels an hour. The size of the cylinder sheller is denoted by the manufacturer's number only. Cylinder shellers have a large capacity ranging up to 800 bushels an hour for the largest sizes.

330. Selection of a sheller.—The following are the requisites for a good portable corn sheller. First and probably the most important feature to look to is the frame. This should be made very strong. It should be mortised and tenoned and secured together by means of rods or bolts. The wood should be either of ash or oak. The bearings for all parts where there is considerable power placed upon them should be long, well secured to the frame, and, where possible, made dust proof. They should also be supplied with plugs or oil cups to keep all grit and dust from entering. The feeding shaft should be strong, and the lugs should be of chilled cast iron or cast steel. The feeder box should be supplied with agitators to prevent the corn piling up at the lower end and thus allowing the sheller to run partially empty. For large job work the machine should be provided with a drag carrier of length from about 10 to 20 feet. Where the cribs are extra long it is well to have two sections of about this length. The rag irons should be separate as well as a combined adjustment. The sheller should be so constructed that it will not injure it to throw the feeder box and the feeder bar into operation while running. On either side of the sheller there should be an attachment for a grain elevator. The mechanism for receiving the power should be so constructed that the power, if necessary, can be applied upon either side. The cob carrier should be of the swing type with long enough lugs on the chain and velocity enough to convey the cobs away without allowing them to choke at the base. In the sheller there should be plenty of surface for the cobs to pass over so the corn can all separate from them.

In selecting a corn sheller and making the first trial, do not condemn the machine if it requires a large amount of power to run it. Possibly the fault is not in the sheller, but is in the condition of the corn. Corn which is green or damp requires very nearly, if not altogether, twice the power to shell it that dry corn requires.

CHAPTER XII

FEED MILLS

331. Development.—The mill was one of the first inventions of man. Feeding of cracked or broken grain to domestic animals has been practiced for many years; however, the practice did not become general until the introduction of the portable mill. The first mills were equipped with stone buhrs, but metallic plates were made use of at a very early date, for they have been mentioned in history. A description of a French mill using metallic buhrs is at hand which was used to grind grain for the soldiers in the army of Napoleon I.

332. Buhrs and plates.—The grinding depends largely upon the buhrs or plates. They are the parts which do the actual grinding; receiving the whole grain, they gradually reduce it to a meal.

The **stone buhr** is used to some extent to-day where a fine meal is desired. The meal from stone buhrs may be used for human food. Buhr stones must have a cellular structure to prevent them from taking on a polish and give them a better grip for grinding. The buhr stone must also be very tough. The best are imported and are known as French buhrs. Good buhr stones are quarried at Esopus, New York, and practically all of the buhr stones used in the United States come from this place. The buhr stone usually has a wrought-iron band shrunk over it to strengthen it. It must be sharpened with a chisel when worn, hence it is not popular for small farms.

Metallic buhrs.—Nearly all of the plates used on farm

mills or grinders are made of chilled iron, though tool steel and bronze are used to some extent.

Chilled iron plates or **buhrs** vary in shape, the usual form being two flat disks which are provided with ribs or corrugations to carry the grain to the outer edge between the milling surfaces (Fig. 175). The **cone buhr** is the result of an attempt to increase capacity by increasing the surface.

The **steel buhr** is made in the shape of a roller with a milled surface. The roller mills as used in flouring mills

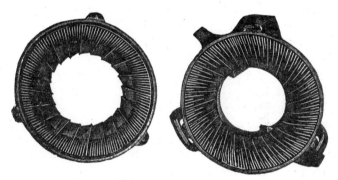

FIG. 175—CHILLED IRON BUHRS

are not used in preparing feed for stock to any extent. It is stated that the steel buhr has a large capacity, but will fill or clog when damp grain is being ground.

The **duplex buhr** has two grinding surfaces. The moving plate moves between two stationary plates (Fig. 178).

In order to grind ear corn a crusher is often provided to reduce the ears to pieces small enough to be fed to the buhrs. In sweep mills the crushing teeth are made a part of the main buhrs.

333. Sweep mills.—The simple sweep mill consists of two conical buhrs. The inner one remains stationary, while the outer is rotated by a sweep. Nearly all sweep mills are arranged to grind ear corn. Fig. 177 illustrates a common type of the sweep mill. In order to increase

FIG. 176—BUHRS FOR A SWEEP MILL

the capacity of the mill one of the buhrs is geared up until it makes 3, or even 9 to 11, revolutions for each round of the team.

FIG. 177—A SWEEP MILL

334. The hitch.—The usual arrangement with the simple sweep mill is to hitch the team to the end of the single sweep. Some makers arrange to hitch the horses

tandem, the claim being that the work is more evenly divided between them, as they work upon an equalizer and each horse travels in the same circle.

With triple-geared or higher-geared sweep mills the capacity for grinding is so great that two horses are not sufficient to furnish the power; more horses must be added. The horses may be hitched in teams to sweeps opposite each other with an equalizer across or placed in tandem, as referred to.

335. Combination mills, or mills in combination with a small sweep power, are manufactured to enable the owner to drive other machinery such as a corn sheller. Such a mill is confined to the geared sweep type.

POWER MILLS

336. Power mills are operated by belt or tumbling rod. Following is a discussion of the important parts of power mills.

A **balance wheel** is sometimes placed upon a mill to prevent the mill from choking due to an extra demand for power which will occur at times. The balance wheel is considered a good thing to have on a mill.

Divided hopper.—It is often desired to grind at least two kinds of grain at a time. To accomplish this a divided hopper is provided.

Safety device.—It often occurs that some hard substance, as a nail or a nut, becomes mixed in the grain and is placed in the mill. The safety device is a wooden break pin or spring catch, which permits the buhrs to open without damaging the mill.

The **quick release** is for the same purpose as the safety device, but is operated by hand. By its use the machine may be prevented from clogging when heavily loaded for any reason.

337. Sacking elevators.—When desired, all larger machines may be obtained with a sacking elevator, provided with a divided spout, to which two sacks may be attached at a time. While one sack is filling, the other may be removed and an empty sack adjusted in its place.

338. The selection of a feed mill.—Feed mills for farm purposes should have their frames constructed of cast

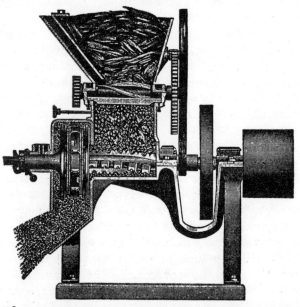

FIG. 178—A SECTIONAL VIEW OF A POWER MILL WITH DUPLEX BUHRS AND
CRUSHING KNIVES

iron, in such a way that there is no binding in the bearings and all bearings may be well protected from the dust. The buhrs should have a device to release them when some foreign substance, such as stones, nails, nuts, etc., enters the mills. Besides this safety device there must be another which is handy and will regulate the buhrs in a manner so they may be opened or closed according to the fineness to which the grain is to be ground. The buhrs should be attached to the shaft or mill in such a manner that they will not wobble and thus rub against each other under any condition whatever. This device should also be made substantial enough and accurate enough so the buhrs can be adjusted to almost any fineness and not interfere with each other. In a corn and cob grinder which is driven by a belt or tumbling rod, the hopper should be divided and should have a feed regulator so the ear corn and fine grain may be regulated as desired. There should also be a regulating device between the crushing cylinder and grinding buhrs. This is quite often effected by means of a lever and vibrating shutter, the former receiving its motion from the main shaft of the mill.

339. Alfalfa mills are used in reducing alfalfa hay to meal suitable for poultry and other stock. The mill has a cutter which cuts the hay into short lengths before passing to the buhrs. Alfalfa may be ground in the corn mill if the hay is passed through a hay cutter first. To grind successfully, alfalfa hay should be very dry. The capacity of alfalfa mills varies from 50 to 100 pounds of ground alfalfa an hour for each horse power used.

340. Capacity of feed mills.—The amount of feed ground an hour depends largely upon the degree of fineness of the ground meal and the condition of the grain as to moisture. It is to be expected that a mill with new sharp buhrs will have a much larger capacity than a mill with worn buhrs. Where a good quality of meal is produced a mill should be expected to grind at least four to five bushels of corn, or two to three bushels of oats an hour for each horse power used. Grinding ear corn the capacity will be one-third less.

341. Corn crushers.—It is within only the past three or four years that the value of crushed corn has become generally known to the cattle feeders. One principal reason for this is that in crushing corn the crushers may be so arranged that the husks may be chopped with the ear. By this means the feeder is enabled to give his cattle snapped corn which is broken or crushed fine enough so it is practically a coarse shelled corn mixed with ground cob and husks. One great advantage derived from such a scheme is that the crushing of the corn can be done very cheaply, it requiring only two or three horse power to crush 40 or 50 bushels an hour. Several feed grinders for grinding corn and cob are provided with a separate crusher and it is a question if this is not the most profitable means of grinding the corn and cob.

CHAPTER XIII

WAGONS, BUGGIES, AND SLEDS

342. Development.—Carts and wagons were used at a very early date, for in the Book of Genesis we find that when Pharaoh advanced Joseph to the second place, "he made him to ride in the second chariot he had." The chariot is only a form of cart. Later in Joseph's time we find that he sent wagons out of the land of Egypt to convey Jacob and his whole family to the land of his adoption. Not only did they have wagons and chariots at a very early date, but they were of similar construction to those of the present, for in the Book of Kings we read, "And the work of the wheels was like the work of a chariot wheel; their axletrees, and their naves, and their felloes, and their spokes were all molten." It is not known just when wheels were first bound with tires of iron, a practice which is of the greatest importance in the construction of the wheel. Wooden wheels without tires have been used in some countries until quite recently, and good authority states that they have a limited use to-day.

The use of carriages for general purposes began in the eighteenth century, though steel springs were introduced as early as the fourteenth. In 1804 Obadiah Elliott invented the elliptical spring. It was early in the nineteenth century that the greatest development took place. During this period Telford and Macadam were able to establish a system of good roads in England.

Carts for the hauling of loads are used to some extent in European countries and to a very limited extent in the

United States. Their use in the Middle West, however, is very rare. The general use of teams and the advantages of the wagon for larger loads are responsible for this.

WAGONS

The essential features of a farm wagon are durability, convenience, lightness of weight and draft. These feat-

FIG. 179—A MEXICAN CART OF 1865. IMPORTED IN 1883 BY MESSRS. SCHUTTLER AND HUTZ OF CHICAGO, AND DONATED LATER TO THE SMITHSONIAN INSTITUTION, WASHINGTON, D. C.

ures depend upon the material, workmanship, and construction used in building the wagon.

343. Material.—Perhaps there is no service to which material may be placed which is as exacting and as severe

FIG. 180—A MODERN FARM WAGON WITH BOX BRAKE

as that required of material used in the construction of
wagons and buggies. All wood should be carefully selected and thoroughly dried both in air and in kiln.
Well-seasoned black birch is probably best for hubs;
best-seasoned white oak for spokes, felloes, bolsters,
sandboards, and hounds; hickory is preferable for axles,

FIG. 181—A SECTION OF A WAGON HUB SHOWING THREE METHODS OF
FORMING THE SPOKE SHOULDERS. THE ROUND SHOULDERS ARE
SAID TO BE MUCH STRONGER AND MORE DURABLE

although the best straight-grained white oak is good. All
metal parts should be of good Norway iron or mild steel.

344. **Wheels.**—All wooden wheels should be dished or
the outer face of the wheel should present a concave surface. The dish in the wheel makes it much stronger,
which may be illustrated with a paper disk and a paper
cone. The cone is much stiffer. For front wheels this
dish should be from ⅜ inch to ⅝ inch, and for rear
wheels from ½ to ¾ inch. At one time, wheels were
given much more dish than at present. An English
writer states that cart wheels should be dished as much
as 3 inches. By giving the wheels an excessive amount

of dish, the cart bed may be made much wider. It does not matter greatly whether the felloes are bent or sawed, as the merits of the two methods are about equal. A rivet should be placed on the side of each spoke to prevent splitting. The felloes should be well doweled and the tire bolted to them. The standard height of wheels for a farm wagon with 3-inch skein or over is 3 feet 8 inches for the front wheels, and 4 feet 6 inches for the rear wheels. Smaller wagons have wheels of less

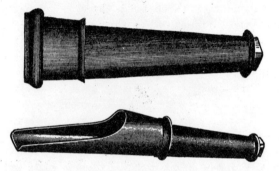

FIG. 182—THE UPPER IS THE CAST; THE LOWER, THE STEEL WAGON SKEIN

height. There is a tendency to use wheels of smaller diameter when wide tires are used. The thickness of the tire varies from ½ inch to ¾ inch.

345. **The axles** should have as few holes in them as possible. Clips can nearly always be used instead of bolts excepting for the king bolt. A well-secured truss rod should be placed beneath each axle, and it is better if it is secured to the skeins.

The skein may be of either cast iron or steel. In level countries the former is preferable, while among the hills and mountains the latter with a long sleeve is probably more serviceable. Skeins should have a large throat to take in all the wood possible, since this is the weakest point in the axle. They should gradually taper towards the nut so they can be forced on perfectly tight and not have to be bolted, as this weakens the axle.

346. Gather.—In setting the skeins the under side should be nearly parallel with the ground and the center of the nut end should be a trifle farther forward than the shoulder. The former is called **bottom gather** and the latter **front gather.** This is so that the wheel will not have a tendency to run towards the nut, to overcome the inclination of the dish of the wheel and keep the box rubbing against the collar of the skein. If the front edges of the felloes are ½ inch closer together than the back, it is sufficient.

347. Tire setting is possibly the most important part of wagon making, since the wheels invariably give out long before any other part. In purchasing a new wagon, it is difficult to tell whether the tires are properly set. However, always avoid buying wheels that have more or less dish than stated above. When having tires reset, see that the smith cuts enough out of the felloe to allow it to draw up snugly on to the spokes and force the spokes into the hub perfectly. Do not allow him to cut out so much that when the felloe is drawn together the wheel is dished more than stated above. Should he not cut out enough of the felloe to accomplish the tightness just stated the wheel will be known as felloe bound and it will be only a short time until the spokes will rattle in the rim or squeak at the hub.

348. The reach in itself is not such an important part, as any person can soon supply a new one. However, the way it is connected to the front axle and passes through the rear is very vital, since it will soon chafe in these places and eventually ruin the gears. See that there is a plate on the under side of the sandboard and on top of the front axle, also see that there is a metal sleeve for the reach to pass through between the rear axle and bolster.

349. Tongue, neckyoke and whiffletrees are all essential, but not so important in their construction. They should all be made of the best selected oak except the doubletrees, which should be of hickory. Wherever there is any wear there should be metal plates or collars. It is well that the tongue be reënforced by an iron strip beneath and that the pole cap have an extra kink in front of the neckyoke lock to prevent the neckyoke from slipping off.

350. Other parts.—The same may be said of sandboards and bolsters as of axles. Between sandboard and bolster there should be a cup and cone plate with flanges which extend over the sides to prevent splitting. On top of each bolster there should be a plate of metal. The king bolt should have a large, flat head to prevent cutting into the bolster.

It does not matter so much as to the length and shape of the hounds, as it does to their bracing and fastening to the axles. Therefore see that they are well braced and so securely fastened that they will not work loose and soon wear at that point.

351. Wide and narrow track.—Two widths of tracks are in general use in the United States. The narrow track measures 4 feet 6 inches center to center of tires on the ground. The wide track is 5 feet measured in the same way. Although the use of each track is confined to certain sections, it results in much inconvenience at the borders of the districts where both styles are used. It is necessary to specify the width of track when purchasing a vehicle of any sort.

352. The box.—The wagon independent of the box is often spoken of as the gear. The box, or what is sometimes called the bed, may be removed and a hay rack or the gear may be used independently for the hauling of logs or lumber. The box of a narrow-track farm wagon is found to be the most convenient when it is 3 feet wide and 10 feet long inside, and made up of three sections, 14, 12, 10 inches deep. The second is spoken of as the top box and the third as the tiptop box. A box of the above dimensions will hold approximately two bushels for each inch in depth. A box of this size requires 3 feet 2 inches between the standards on the bolster, and is 10 feet 6 inches long outside. The sides of the box should be of the best selected yellow poplar and the bottom of 3-inch quarter-sawed yellow pine flooring with oak strips on the under side. A metal plate should be riveted on where the bolster rubs, and a rub iron of good design and secure attachment should be placed where the front wheel rubs. A device should be provided to hold the box sections securely together.

353. Brakes.—Wagon brakes are required in hilly localities. Two general types of wagon brakes are in use, the box brake or the brake attached to the wagon box, and the gear brake, attached to gear independent of the box, except that a lever attached to it is provided to be used when the box is used. The gear brake has two advantages in that it does not weaken or injure the box in any way, when used, and it may be used when the gear is used without the box. The box brake has a tendency to chatter and loosen the floor of the box.

354. Painting.—All of the wooden parts of the gears should be boiled in linseed oil and then one coat of paint applied before the ironing is done. The former process drives all moisture from the wood and fills the pores so the paint adheres well; the latter keeps moisture from entering, thus preventing the wood from rotting under the iron. After ironing, two more coats of red lead paint should be added, then stripes, and finally a coat of wagon varnish. The box should be sandpapered, then painted with three coats of good pigment, after which it is striped and varnished.

355. Capacity.—As a wagon is subjected to shocks, it must be designed to carry many times any load which may be placed upon it. The following table is the average capacity of wagons as furnished by several manufacturers:

Wagons with Skeins		With Steel Axles	
Size of Skein	Capacity	Size of Axle	Capacity
$2\frac{1}{8}$	800	$1\frac{1}{16}$	600
$2\frac{1}{4}$	1,000	$1\frac{1}{8}$	800
$2\frac{3}{8}$	1,200	$1\frac{1}{4}$	1,200
$2\frac{1}{2}$	1,600	$1\frac{3}{8}$	1,600
$2\frac{3}{4}$	2,000	$1\frac{1}{2}$	2,250
3	3,000	$1\frac{5}{8}$	3,000
$3\frac{1}{4}$	4,000	$1\frac{3}{4}$	4,000
$3\frac{1}{2}$	5,500	2	5,000
$3\frac{3}{4}$	6,500	$2\frac{1}{4}$	6,500
4	8,500	$2\frac{1}{2}$	9,000
$4\frac{1}{4}$	10,000	3	15,000
$4\frac{1}{2}$	12,000		

356. Draft of wagon.—The draft of a wagon is the resistance encountered in moving the wagon with its load. It is often called tractive resistance, and is worthy of careful consideration, for a reduction in the draft of wagons not only means increased efficiency on the part of the draft animals, but also a reduction in the cost of transportation. The draft of wagons is made up of three elements: (a) axle friction, (b) rolling resistance, and (c) grade resistance.

357. Axle friction is the resistance of the wheel turning about its axle similar to the resistance of a journal turning in its bearing, independent of the other elements of draft. Axle friction is usually a small part of the total draft. The power required to overcome it diminishes as the ratio between the diameters of the wheel and axle increases. Thus in Fig. 183 if R be the radius of the wheel, r the radius of the axle, from the principle of the wheel and axle—

$$\text{Power : Axle friction : : } r : R$$
$$\text{Power} = \frac{\text{Axle friction}}{R/r}$$

In the standard farm wagon R/r has a value of from 11 to 20, or an average of about 15.

Morin found in his experiments, which have been considered a standard for years, that with cast-iron axles in cast-iron bearings lubricated with lard, oil of olives or tallow gave a coefficient of friction of 0.07 to 0.08 when the lubrication was renewed in the usual way. Assuming 0.08 to be the coefficient of friction and 15 to be the ratio between wheel and axle diameters, the force re-

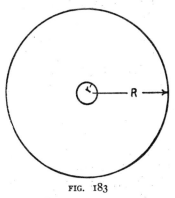

FIG. 183

quired per ton to overcome friction would be between 10 and 11 pounds. Another authority* states that the tractive power required to overcome axle friction in a truck wagon which has medium-sized wheels and axles is about 3½ to 4½ pounds a ton. The use of ball and roller

*I. O. Baker, "Roads and Pavements."

bearings would tend to reduce the axle friction and manufacturers trying to introduce these bearings claim a great reduction in draft. No doubt there are other advantages in the use of ball and roller bearings beside a reduction in draft. It is not thought that the dished wheel and bent axle are of a construction that tends to reduce axle friction to a minimum. It is hoped that experiments will be conducted at an early date to determine accurately the axle friction of wagons.

358. Rolling resistance.—Rolling resistance corresponds to rolling friction in that it is due to the indentation or cutting of the wheel into the road surface, which really causes the wheel to be rolling up an inclination or grade. The softer the road bed the farther the wheel will sink into it, and hence the steeper the inclination. The height of wheel influences the rolling resistance in that a wheel of large diameter will pass over an obstruction with less power, as the time in which the load is lifted is lengthened. There is also a less tendency upon the part of a large wheel to cut into the surface, due to the larger area presented at the bottom of the wheel to carry the load. Elaborate experiments have been conducted by T. I. Mairs, of the Missouri experiment station, in regard to the influence of height of wheel upon draft of wagons. Three sets of wheels were used with six-inch tires and a net load of 2,000 pounds was used in all cases. The total load for the high wheels was 3,762 pounds, for the medium wheels 3,580, and for the low wheels 3,362.

The high wheels were 44-inch front wheels and 56-inch hind wheels.
" medium " " 36 " " " " 40 " " "
" low " " 24 " " " " 28 " " "

EFFECT OF HEIGHT OF WHEELS ON DRAFT *
Draft in Pounds per Ton

Description of Road Surface	High Wheels	Medium Wheels	Low Wheels
Macadam; slightly worn, clean, fair condition..................................	57	61	70
Dry gravel road; sand 1 inch deep, some loose stones............................	84	90	110
Earth road—dry and hard.................	69	75	99
" " —thawing ½-inch sticky mud.	101	119	139
Timothy and bluegrass sod, dry, grass cut.	132	145	179
" " " wet and spongy....	173	203	281
Corn; field flat culture across rows, dry on top..................................	178	201	265
Plowed ground, not harrowed dry and cloddy	252	303	374

The **width of tire** also influences the rolling resistance to a great extent. The wide tire on a soft road bed is able to carry the load to better advantage and prevent the wheel cutting in as far as it would otherwise.

The rolling resistance as indicated in the above remarks depends largely upon the condition of the road surface. The harder and smoother the road surface the less will be the rolling resistance. It is for this reason that much larger loads may be hauled upon good hard roads than upon poor soft ones. Prof. J. H. Waters, at the Missouri experiment station, has conducted extended experiments to determine the influence of the width of tire upon the draft of wagons when used on various road surfaces. The wheels used were of standard height and were provided with 1½-inch and 6-inch tires. The summary of the results of these experiments states that the wide tires gave a lighter draft except under the follow-

* Missouri Agricultural Experiment Station, Bulletin No. 52, 1901.

ing conditions: (*a*) When the earth road was muddy, sloppy and sticky but firm underneath, (*b*) when the mud was deep and adhered to the wheels, (*c*) when the road was covered with deep loose dust, and (*d*) when the road was badly rutted with the narrow tire.

INFLUENCE OF WIDTH OF TIRE UPON DRAFT*

Draft in Pounds per Ton

Description of Road Surface	Width of Tire	
	1½-Inch	6-Inch
Broken stone road—hard, smooth, and no dust.....	121	98
Gravel road—hard and smooth	182	134
" " —wet, loose sand, 1 to 2½ inches deep.	246	254
Earth road loam, dry dust, 2 to 3 inches deep......	90	106
" " " dry and hard, no dust...........	149	109
" " " stiff mud, dry on top, spongy underneath...................	497	307
" " clay, sloppy mud, 3 to 4 inches hard below	286	406
" " clay, stiff, deep mud.................	825	551
Plowed land harrowed smooth and compact........	466	323

Besides the reduction of draft attained in the majority of cases with the use of wide tires, there is another important advantage from their use, as there is less tendency to rut and destroy the road surface. It is believed that this feature should be placed before all others.

There is a slight increase in draft with an increase in speed. Morin, who conducted experiments to determine the relation between draft and speed, found that the draft increased about as the fourth root of the speed. The draft upon starting a load is greater than after motion has been attained, and is due to the settling of the load into the road bed, the increased axle friction of rest, and

* Missouri Agricultural Experiment Station, Bulletin No. 39, 1897.

the extra force required to accelerate the load. Springs tend to reduce draft, as they reduce the shocks and concussions due to the unevenness and irregularities of the road surface. Their effect is greater at high speeds than at lower.

359. Grade resistance.—Grade resistance involves the principle of the inclined plane, and may be explained as the force required to prevent the load from rolling down the slope. It is independent of everything except the angle of inclination.

In Fig. 184 if W be the load and P the grade resistance, AB the height of the grade and CB the length, by completing the force diagram similar triangles are obtained, from which it is seen:

$$P : AB :: W : AC, \quad \text{or} \quad P = W \times \frac{AB}{AC}$$

As AC is very nearly equal to BC for ordinary grades, no great error will be accrued by substituting BC for AC. Grades are usually expressed in the number of feet rise

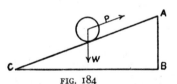

FIG. 184

and fall in 100 feet, or in the number of per cent the total rise is of the length of the grade. Then for practical purposes the grade resistance is equal to the per cent of the total load, which expresses the grade. For example, if the grade is 5 per cent and the load 2,000 pounds, the grade resistance will be 100 pounds. The foregoing analysis does not take into account the way the load is placed on the wagon or angle of hitch, which may lead to error.

360. Handy wagons.—The name handy wagon is given to a low-wheeled, broad-tired wagon used about the farm for hauling implements, grain, and stock. They are used

to a limited extent in road transportation. Two styles
of wheels are used, the metal with spokes cast in the hub
and riveted into the tire, and a solid wooden wheel bound
with a tire and provided with a cast hub.

The metal wheel may be had in any height from 24
inches up. The wheel with staggard oval spokes is considered stronger than the straight spoke wheel, as it is
able to resist side hill stresses to better advantage.

The solid wooden wheel is very strong and there is no
tendency for the wheel to fill with mud above the tire.
The fact that the wheel proper is made of wood requires
an occasional setting of tires, but this is not often, as the
wheel is filled with circular wooden disks with the grain
of the sections at right angles, and there is little shrink-
age on account of the small diameter of the wheel. Four-
or 5-inch tires are common widths used on handy wagons, although almost any width may be obtained.

Some handy wagons are made very cheaply and sold at
a very low price. These wagons are poorly ironed, do
not have any front or rear hounds, and are poorly finished. Others are made with as much care as the standard farm wagon and are as well finished. Care should
be used in the selection of a handy wagon. Although
boxes may be used upon handy wagons the wagon used
about the farm is usually equipped with a rack or a flat
top which readily permits the loading of implements,
fodder, etc.

BUGGIES AND CARRIAGES

361. Selection.—Light vehicles for driving have been
in use since the introduction of springs and good roads.
The points which make a buggy or a carriage popular are
lightness, neatness of design, excellent and durable finish, good bracing, a reliable fifth wheel, well-secured

clips, and a body sufficiently braced and stayed and, if
so provided, with a neat leather or at least leather quarter top.   Leather quarter is the name given to tops made
with leather sides above the curtains, while the roof is
made of the cheaper material, rubber or oil cloth.

It is very hard to detect quality in a buggy and the re-
liability and guarantee of the manufacturer must be depended upon to a large extent.   As in the construction
of wagons and implements, poor quality may be detected
by poor workmanship used in the construction.   Only
the best materials, carefully cured, should be used in the
construction.   The wheels and other wood parts of the

FIG. 185—A LONG-DISTANCE BUGGY AXLE.   NOTE THE PROVISION MADE TO
EXCLUDE DUST AND DIRT

gear should be made of best hickory.   This is especially
true of the wheels, which must meet with very hard
service.   The rims of the wheels should be well clipped
and screwed.

362. **The body or box** should be made of the very best
yellow poplar and should be well screwed and braced.
The plain top buggy has two common styles of bodies:
the **piano box,** which is narrow and has the same height
of panel all around, and the **corning body,** which has low
panels just back of the **dashboard.**

363. **Hubs.**—Two styles of hubs are in general use,
the **compressed** hub with staggard spokes and the **Sarven
patent** hub.   The former is perhaps the stronger but
more difficult to repair.

There are many other parts which might be **men-**

tioned, as the styles of springs, spring bars, box loops, etc., but it is not deemed wise to take up space.

364. The painting of a buggy is of great importance and should be done only by an expert. Several coats of filler should be used, and between coats it should be

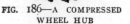

FIG. 186—A COMPRESSED   FIG. 187—A SARVEN
     WHEEL HUB   WHEEL HUB

well sandpapered. In all, there should be 20 to 24 coats applied. It is stated that the varnish for the body should be first-grade copal, and for the gears second-grade copal, which should be very carefully rubbed between coats and the final coat should be rubbed with the palm of the hand.

SLEDS

365. Utility and selection.—Sleds were the first means of conveyance known to man, and among the uncivilized they are still the only conveyance. There has probably been as great a change made in the sled as in the wagon since man commenced to improve his machinery.

Due to the variety of work required of sleds and the climatic conditions, there is almost invariably a different type of sled required in every locality. In heavily timbered countries where there is an extended season of snow, sleds are made with as much care as wagons, while in communities where sleds are used only at intermittent times of the year and then only as a substitute for a wagon with light loads, they are very much more cheaply built.

Where the **runners** of a sled are bent they should be of either ash or hickory. If the natural curve of a tree is used, good hard wood will do. If the curve is sawed, white oak is better. All other parts should be of oak.

The **knees** should be fastened by means of two bolts on each end. This will prevent splitting. All connections are better if made flexible, and it is more convenient to have the front bob connected so it can turn under the load. The **shoes** are more economical when made of cast iron and removable. In communities where there is no continued season of snow a cheaper type of sled is sufficient. In such cases the shoes can be made of wrought iron, the bobs connected directly by a short reach and eyes, and the flexible parts dispensed with.

366. Capacity.—A bob sled with two knees in each bob ought to have a capacity of about 4,000 pounds, and one with three knees, of 6,000 pounds.

There is practically no limit to the load a team can handle on a sled provided they can start it. In most cases it is better to carry a bar to assist in starting the load and thus avoid the troublesome lead team.

In hilly countries it is essential to have some method for holding the load back in descending and to keep it standing while the team breathes upon ascending a hill. A short chain attached to the runner and dropped beneath it will hold the load back when descending a hill. In some localities a curved spike extending to the rear is bolted to the sled in such a manner as to prevent the sled from sliding backward when pressed to the snow by the teamster.

CHAPTER XIV

PUMPING MACHINERY

367. Early methods of raising water.—The oldest method of raising water was by bailing. The vessel and the water it contained were raised either by hand or by machines to which power might be applied. The buckets were provided with a handle or a rope when it was desired to draw water from some depth. To aid in drawing water from wells, the long sweep or lever weighted at one end was devised. This sweep is often seen illustrated in pictures of an old homestead and similar pictures. Following the sweep, a rope over a pulley with two buckets, one at each end, was used. Later, one bucket was used and the rope carried over a guide pulley and wound around a drum. This latter method of raising has not entirely disappeared and is still in use in many places.

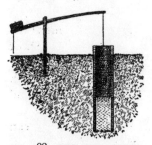

FIG. 188—THE WELL SWEEP, AN OLD METHOD OF RAISING WATER

For raising water short distances and in large quantities, swinging scoops and flash wheels are used. The scoop is provided with a handle and is swung by a cord long enough to permit it to be dipped into the water. The water is simply pitched to a higher elevation much like grain is elevated. Flash wheels are the reverse of the undershot water wheel; the paddles or blades ascending a chase or waterway carry the water along with them. If operated by hand the paddles are hinged like valves and are rocked back and forth in the waterway. Flash wheels are used extensively in Holland in draining low lands.

The Chinese devised at a very early time scoop wheels which have buckets on the periphery. These buckets dip into water and are set at such an incline that they carry almost their full capacity to the upper side, and there they pour their contents into a trough. They are sometimes hinged and are made to discharge their contents by striking against a suitable guide. Wheels of this nature may now be used profitably where a large quantity of water is to be elevated for only short distances.

One of the oldest water-raising devices made famous by history is the Archimedean screw. It consists essentially of a tube wound spirally around an inclined shaft and taking part in the rotation of this shaft. The pitch of the screw and the inclination of the shaft are so chosen that a portion of each turn will always slope downward and form a pocket. A certain quantity of water will be carried up the screw in these pockets as it is rotated. At the upper end of the inclined screw the water is discharged from the open end of the tube.

368. Reciprocating pumps.—As advancement came along other lines of machinery, the early devices for raising water gave way to the introduction of more efficient machines to which may properly be given the name of pumps, the most common of which is the reciprocating pump. A reciprocating pump consists essentially of a cylinder and a closely fitting piston.

369. Classes.—Reciprocating pumps may be divided into two classes:

1. Pumps having solid pistons or plunger pumps.

2. Pumps having valves in the piston or bucket pumps.

Plunger pumps will not be considered in this discussion, for, at the present time, their use is confined almost entirely to steam and large power pumps. Pumps used for agricultural purposes are almost universally of the latter type.

Pumps may further be divided into two distinct classes:

　1. Suction or lift pumps.

　2. Force pumps.

Suction pumps do not elevate the water above the pump standard. The pump standard is the part which is above the well platform when speaking of pumps for hand or windmill power. A pump will then necessarily include the standard, cylinder, and pipes.

370. Pump principles.—Before continuing the discussion it will be well to take some of the principles connected with the action of pumps. The action of a plain suction pump when set in operation is to create a vacuum, and atmospheric pressure when the lower end of the suction pipe is immersed in water causes the vacuum to be filled. Atmospheric pressure amounts to about 14.7 pounds per square inch. Water gives a pressure of .434 pound per square inch for each foot of depth, or each foot of head, as it is usually spoken of. Thus atmospheric pressure will sustain a water column only about 33.9 feet, above which a vacuum will be formed. Pumps will not draw water satisfactorily by suction more than 25 feet, and it is much preferred to have the distance less than 20 feet. It is often an advantage to have the cylinder submerged.

371. Hydraulic information.—The following information will be useful in making calculations involving pumping machinery:

A United States gallon contains 231 cubic inches.
A cubic foot of water weighs 62.5 pounds.
A gallon of water weighs 8⅓ pounds.
A cubic foot contains approximately 7½ gallons.

The pressure of a column of water is equal to its height multiplied by .434. Approximately. the pressure is equal to one-half of the height of water column or head.

Formulas for pump capacity and power:

D = diameter of pump cylinder in inches.

N = number of strokes per minute.

H = total height water is elevated, figuring from the surface of suction water to highest point of discharge.

S = length of stroke in inches.

Q = quantity of water in gallons raised per minute.

$D^2 \times .7854 \times S$ = capacity of pump in cubic inches per stroke.

$$\frac{D^2 \times S}{294} = \text{capacity of pump per stroke in gallons.}$$

$$\frac{D^2 \times S}{35.266} = \text{capacity of pump per stroke in pounds of water.}$$

$$\frac{D^2 \times S \times N}{294} = \text{capacity of pump per minute in gallons.}$$

$$\frac{D^2 \times S \times H \times N}{35.268} = \text{number of foot-pounds of work per minute.}$$

A rule which may be used to calculate roughly the capacity of a pump is as follows: The number of gallons pumped per minute by a pump with a 10-inch stroke at 30 strokes per minute is equal to the square of the diameter of the cylinder in inches. From this rule it is easy to calculate the capacity of a pump of a longer or shorter stroke and making more or less strokes per minute.

372. Friction of pumps.—Pumps used to pump water from wells are of rather low efficiency; on an average, 35 per cent of the power is required to overcome friction

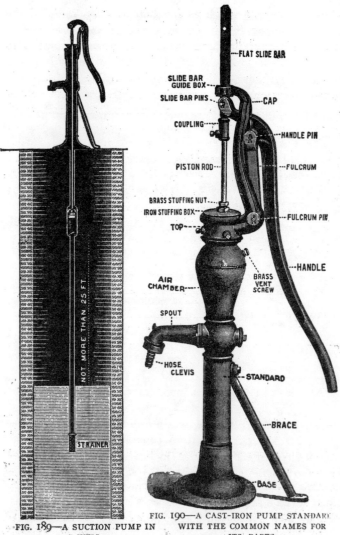

FIG. 189—A SUCTION PUMP IN A WELL

FIG. 190—A CAST-IRON PUMP STANDARD WITH THE COMMON NAMES FOR ITS PARTS

alone. Often as much as one-half or even more of the power is required for this purpose. A common rule in use to determine approximately the power required to operate a farm pump is that one horse power is required to lift 30 gallons 100 feet per minute. From this rule it is easy to calculate for different capacities at more or less head. The rule assumes a mechanical efficiency of 68 per cent on the part of the pump.

The friction of water flowing in pipes is also very great. The loss of head due to friction is proportional to the length of the pipe and varies about as the square of the velocity of the flow. It is greatly increased by angles, valves, roughness, and obstructions in the pipe.

The following table given by Henry N. Ogden indicates the loss of head due to friction in pipes:

LOSS OF HEAD DUE TO FRICTION*

Flow in Gallons per Minute	Loss of Head by Friction in each 100 Feet of Length	
	½-Inch Pipe	1-Inch Pipe
0.5	4	....
1.0	7	0.3
2.0	17	0.7
4.0	54	1.6
7.0	140	5.3
10.0	224	9.3

The importance of choosing a pipe of sufficient size for the flow per minute and the length of pipe is shown by this table. For instance, suppose it is desired to deliver seven gallons a minute at a distance of 500 feet. The ½-inch pipe would require an impractical head of

* The Installation of Farm Water Supplies—Cyclopedia of American Agriculture, Vol. I., page 294.

700 feet, while 1-inch pipe would need only about 26 feet of head to secure the desired flow.

373. Wells.—The type of pump used will often depend upon the kind of well. Wells are divided into four classes: (a) dug or bored wells, (b) driven wells, (c) tubular wells, and (d) drilled wells. Dug wells are those from which the earth is removed by a bucket, rope, and windlass. These wells are either walled with stone or brick or cased with wooden or tile curbing. Bored wells belong to the same class except the earth is removed from the well with an auger. Pumps for dug or bored wells are independent of the casing, and any common type may be used provided the cylinder is placed within the proper distance of the water. Driven wells are made by attaching a point with a screened opening to permit of a flow of water to the casing, usually 1¼-inch galvanized pipe, and the whole driven to sand or gravel strata bearing water. A driven well does not extend through rock strata. Tubular wells are made by attaching a cutting edge to the well casing, which is usually made of pipe 2 inches in diameter, and which is sunk into the opening made by a drill which operates inside of the casing. The earth and chips of stone are removed by a stream of water which flows out through the hollow drill rod in the form of a thin mud. A screened sand point similar to those used in driven wells is placed in the bottom of the well after it has been finished. A turned flange is provided which prevents the point from passing beyond the casing. A pit 6 feet deep and 4 feet square, walled with brick, stone, or cement, should be placed around driven and tubular wells to permit of the use of underground pumps, or to provide a vent hole to prevent water freezing in the pump standard during cold weather. It is an advantage to have the well at least 6 inches from one side of the pit wall, as this will permit the use of pipe tools to better advantage.

Drilled wells are much like tubular wells except that they are larger, usually 6 or 8 inches in diameter, cased with wrought-iron pipe or galvanized-iron tubing. The pump is independent of the casing and may be removed without molesting it in any way.

Pump cylinders or barrels usually form a section of the casing in driven and tubular wells. The lower check valve is seated below the barrel by expanding a rubber bush against the walls of the well casing in such a way as to hold it firmly in place. It is to be noted that wooden pump rods should be used for deep-driven and tubular wells, for wooden rods may not only be lighter, but displace a large amount of water, reducing the weight on the pump rod during the up stroke.

374. Wooden pumps.—The first pumps were made of wood, simply bored out smoothly and fitted with a piston. The wood used was either oak, maple, or poplar. Later an iron cylinder was provided for the piston to work in. The better pumps of to-day belonging to this class have porcelain-lined or brass cylinders. These lined cylinders are smoother and are not acted upon by rust. Wooden pumps are nearly all lift pumps and can be used only in shallow wells. The cylinder is fitted in the lower end of the stock and no provision is made for lowering it. Wooden pumps are used with wooden piping, the ends of the pipe being driven into the lower end of the stock so as to form an air-tight joint.

375. Lift pumps.—Lift pumps include all pumps not made to elevate water above the pump standard. For this reason the top of the pump is made open and the pump rod not packed, as is the case in force pumps. Lift pumps, in the cheaper types, are cast in one piece, the handle and top set in one direction, which cannot be changed. Another style of light pump is made in which the lower part of the standard is a piece of wrought-iron pipe. The cast standard has one advantage in cold climates, as it permits warm air from the well to circulate around the pipe where it extends into the standard and prevents freezing to a certain extent.

376. Pump tops.—Pump tops are divided into two classes, known as hand and windmill tops. The former permits the use of hand power only, while with the latter the pump rod is extended so as to permit windmill connection. At least two methods are to be found for fastening the pump top in place: set screws and offset bolts. The latter seem to give the best satisfaction, as they give more surface to support the top and are not apt to work loose from the jerky motion given to the pump handle. Windmill tops should be provided with interchangeable guides or bushes, which may be replaced when worn. This is not important, however, as very little wear comes upon the bushes, the forces being transmitted in a vertical direction only.

377. Spouts.—Spouts are either cast with the pump standard or made detachable. They are styled by the makers plain, siphon or gooseneck, and cock spouts. The object of the siphon spout seems to be the securing of a more even flow of water from the pump. If the pump is a force pump, the spout should be provided with some means of making a hose connection. The cock spout is for this purpose, but a yoke hose connection or clevis may be used for the same purpose with a disk of leather in the place of the regular washer.

378. Bases.—Like the spout, the base may be cast with the rest of the pump standard. However, there are two other types found upon the market: the adjustable and split or ornamental. It is a great advantage in fitting the standard to a driven well to have the base adjustable, doing away with the necessity of cutting the pipe an exact length in order to have the base rest upon the pump platform or having to build the platform to the pump base.

379. Force pumps.—Force pumps are those designed to force water against pressure or into an elevated tank. In order to do this the pump rod must be packed to make it air tight. Force pumps are also provided with an air chamber to prevent shocks on the pump. It is common practice to use the upper part of the pump standard for the air chamber. It has a vent cock or a vent screw to permit the introduction of air when the pump becomes waterlogged. With tubular wells it is an advantage to have a pump standard with a large opening its entire length and a removable cap to permit the withdrawal of the plunger or cylinder. The two most common methods of providing for this are to have the pump caps screwed on and to have the cap and the pump top in one piece. In the latter case the entire top is made air tight by drawing it down on a leather gasket or washer on the top of the standard.

380. Double-pipe pumps or underground force pumps. This class of pump is used where the water is to be forced underground, away from the pump to some tank or reservoir. These pumps are built with either a hand or a windmill top. A two-way cock is provided, manipulated from the platform to send the water either out of the spout above the platform or through the underground pipe. As the piston rod of these pumps has to be packed below the platform where it is not of free access, we find in use a method of packing known as the stuffing-box tube to take the place of the ordinary brass bush.

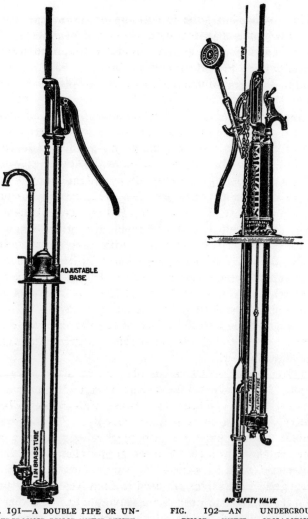

FIG. 191—A DOUBLE PIPE OR UN-
DERGROUND PUMP WITH STUFF-
ING-BOX TUBE AND ADJUSTABLE
BASE

FIG. 192—AN UNDERGROUND
PUMP WITH ORNAMENTAL
BASE AND EQUIPPED WITH A
WINDMILL REGULATOR

The stuffing-box tube is nothing more nor less than an auxiliary piston fitted with the regular leathers. The tube is always made of brass, and does not need attention as often as the regular stuffing box.

381. Pump cylinders.—Three classes of pump cylinders are found upon the market: Iron, brass-lined, and brass-body. Iron cylinders are used mostly in shallow wells. Brass-lined and brass-body cylinders are the most desirable, as they work very smoothly and will not corrode in the least. Iron cylinders are often galvanized to prevent rusting. Brass-body cylinders have the cylindrical portion between the caps made entirely of brass. Brass cylinders are easily damaged by being dented, and when so damaged cannot be repaired to good advantage. Brass being a soft metal, some difficulty is encountered in making a good connection between the cylinder and the caps by screw threads. In order to strengthen the brass-body cylinder at this point, the caps are often fitted on the cylinder by rods at the sides.

Cylinders to be used inside of tubular or drilled wells are made with flush caps to enable a larger cylinder to be put into the well.

382. Valves.—The valves of a pump are a very vital part. Most valves are made of iron in the piston and leather in the cylinder cap. Brass often makes a better valve than iron, as it will not corrode. The valve commonly used is known as a poppet valve, and may have one or three prongs. The single-pronged valve is not interfered with by sand to the same extent as the three-pronged. Ball valves are used in deep-well pumps, but it is very difficult to keep these valves tight. Various materials are used out of which to make the valve seats. One large manufacturer manufactures valve seats of glass and makes many claims for their superiority.

Pump pistons are usually provided with only one cap leather for the piston. For high pressures more are needed, and in the better makes of deep-well pumps the pistons are provided with three or even four leathers.

383. Pump regulators have a hydraulic cylinder attached, into which the pump forces water when the connection with the tank is cut off by a float valve. The hydraulic cylinder is provided with a piston and a stuffing box and a piston rod. Connection is made by a chain to a quadrant on a weighted lever above the platform. This lever is also attached to the pull-out wire of the mill. All the water being forced into the hydraulic cylinder, enough pressure is created to pull the mill out of gear. Safety valves are provided to prevent too great pressures coming on the hydraulic cylinder, which might cause breakage.

384. Chain and bucket pumps.—Chain pumps have the pistons or buckets attached to a chain running over a sprocket wheel at the upper or crank end, and dip in the water at the lower. The buckets are drawn up through a tube, into which they fit and carry along with them the water from the well. The chain pump is suited only for low lifts.

Another type of pump similar to the above and sometimes styled a water elevator has buckets open at one end, attached to the chain. These are filled at the bottom and are carried to the top, where they are emptied. It is claimed the buckets carry air into the water and this has a beneficial effect.

385. Power pumps are not used very extensively about the farm except for irrigation and drainage purposes. When the power is applied with a belt the pump is known as a belted pump. If provided with two cylinders, it is known as **duplex**; if three, **triplex**. The cylinders may

be single or double acting. In double-acting pumps the
water is discharged at each forward and backward stroke.
The capacity of a double-acting pump is twice that of a
single-acting pump. A direct-connected pump is on the
same shaft with the motor or engine, or coupled thereto.

FIG. 193—A ROTARY PUMP

386. **Rotary pumps** are used to some extent in pumping about the farm. They are not suited for high lifts,
as there is too much slippage of the water past the
pistons. They are not very durable, and it is doubtful if
they will ever come into extensive use.

387. **Centrifugal pumps** are used where a large quan-

tity of water is to be moved through a short lift, as in drainage and irrigation work. They are efficient machines

FIG. 194—SECTION OF A ROTARY PUMP SHOWING PISTONS

for low lifts at least, and will handle dirty water better than any other kind of pump. Centrifugal pumps are

FIG. 195—CENTRIFUGAL PUMP

made with either a vertical or a horizontal shaft. The pumps with a vertical shaft are called vertical pumps and may be placed in wells of small diameter. This class of pump gives but little suction and works the best when immersed in the water.

388. The hydraulic ram.—Where a fall of water of sufficient head and volume is at hand, it may be used to elevate a portion of the flow of water to a higher elevation. The action of a hydraulic ram depends upon the intermittent flow of a stream of water whose momentum when brought to rest is used in forcing a smaller stream to higher elevation. The ram consists essentially of (*a*) a drive pipe leading the water from an elevated source to the ram; (*b*) a valve which automatically shuts off the flow of water from the drive pipe through the overflow, after sufficient momentum has been gathered by the water; (*c*) an air chamber in which air is compressed by the moving water in the drive pipe in coming to rest; and (*d*) a discharge pipe of smaller diameter leading to the elevated reservoir.

TABLE OF PROPORTIONATE HEAD, GIVING HIGHEST EFFICIENCY IN OPERATION OF HYDRAULIC RAM*

To Deliver Water to Height of			Place Ram under				Conducted Through		
20 feet above ram			3 feet	Head of Fall			30 feet of Drive Pipe		
40 " " "			5 "	"	"	"	40 "	"	"
80 " " "			10 "	"	"	"	80 "	"	"
120 " " "			17 "	"	"	"	125 "	"	"

Under the foregoing conditions about 12 times as much water will be required to operate ram as will be discharged.

Hydraulic rams are manufactured in sizes to discharge from 1 to 60 gallons a minute, and for larger capacities

* The Gould Company, Chicago.

rams may be used in batteries. To replenish the air in
the air chamber, a snifting valve is placed on the drive
pipe. In freezing weather it is necessary to protect the
ram by housing, and often artificial heat must be supplied.

389. Water storage.—Owing to the fact that water
must in nearly all cases be pumped at certain times which

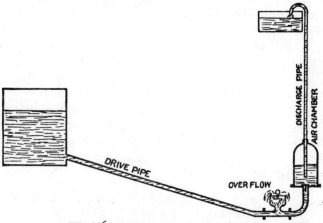

FIG. 196—HYDRAULIC RAM IN OUTLINE

may vary greatly in the intervals between each other,
some form of water storage must be had in order to
secure at all times an adequate supply to meet the constant needs. It is not only necessary to have a supply
to furnish water for stock and household needs, but also
for fire protection.

390. Amount of water needed.—The amount of water
required for household purposes with modern conveniences has been found to be about 20 gallons a person,
large or small. A horse will drink about 7 gallons a
day and a cow 5 to 6 gallons. From this data the amount
of water used a day may be estimated. If a windmill is

used to pump the water, three to four days' supply should be stored to provide for a calm. If a gasoline engine is used, it will not be necessary to store for so long an interval. Two systems of storing water are now in use: the elevated tank and the pneumatic tank.

391. Storage tanks.—*The elevated tank* may be placed outside on a tower, or in the building upon an upper floor. The objection to placing a tank in a building is the great weight to be supported. It has the advantage of being protected from dirt and the weather. The elevated tank on a tower is exposed to freezing in winter and to the heat of the sun in summer. Furthermore, a tower and a wooden tank are not very durable. The elevated tank is cheaper than the pneumatic system where a large amount of storage is desired. A reservoir located on a natural prominence, when such a location can be secured, offers many advantages in the way of capacity and cheapness.

The pneumatic or air-pressure system has an inclosed tank partly filled with air and partly with water. When filled the air is under pressure, and, being elastic, will give the same kind of pressure to the water as an elevated tank. One of the principal advantages of the air-pressure system is that the tank may be buried in the ground or placed in the cellar in a cool place. The disadvantage is a limited capacity for the cost.

If water be pumped into a closed tank until the tank is half full, the air contained will give a pressure of about 15 pounds a square inch, which is sufficient to force the water to a height of 33 feet. Air in the tank follows the well-known law of gases known as Boyle's law—pressure $\times$ volume $=$ constant. If the air be pumped to a pressure of 10 pounds before the introduction of the water, the maximum discharge from the tank will be had at the common working pressures. The water capacity of a tank will not be more in any case than two-thirds the total capacity of the tank. As the water continually dissolves a certain amount of the air, or, rather, carries the air out with it, it is necessary to supply air to the tank from time to time. Pumps are now arranged with an auxiliary air cylinder to supply this air.

It is not advisable to pump air to pressure because it is very slow work, as each cylinderful must be compressed before any is forced into the tank.

Air-pressure tanks must be very carefully made, as air is very hard to contain, much more difficult than steam.

CHAPTER XV

THE VALUE AND CARE OF FARM MACHINERY

392. Value and cost.—Few realize the enormous sums spent annually by the farmers of the United States for machinery. Of the $2,910,138,663, the value of all crops raised in 1899, about 3.4 per cent was spent for machinery. The total amount of money invested in machinery was $749,775,970. The following is the census report of the value of machinery manufactured each census year since 1850:

Year	Total for U. S.	Year	Total for U. S.
1850	$6,842,611	1880	$68,640,486
1860	20,831,904	1890	81,271,651
1870	42,653,500	1900	101,207,428

In closing, it is fitting that the subject of the care of farm machinery be considered, for one reason at least. The American farmers buy each year over $100,000,000 worth of machinery, which is known to be used less efficiently than it should be. The fact that farm machinery is poorly housed may be noticed on every hand. Even the casual observer will agree that if machines were housed and kept in a better state of repair they would last much longer and do more efficient work. It has been stated by conservative men that the average life of the modern binder is less than one-half what it should be.

The care of farm machinery readily divides itself into three heads: First, housing or protecting from the weather; second, repairing; third, painting.

393. Housing.—Many instances are on record where farmers have kept their tools in constant use by good care for more than twice the average life of the machine. The machinery needed to operate the modern farm represents a large investment on the part of the farmer. This should be considered as capital invested and made to realize as large a dividend as possible. The following is a list of the field tools needed on the average 160-acre farm and their approximate value:

1	grain binder	$125.00
1	mower	45.00
1	gang plow	65.00
1	walking plow	14.00
1	riding cultivator	26.00
1	walking cultivator	16.00
1	disk harrow	30.00
1	smoothing harrow	17.00
2	farm wagons	150.00
1	corn planter	42.00
1	seeder	28.00
1	manure spreader	130.00
1	hay loader	65.00
1	hay rake	26.00
1	light road wagon	60.00
1	buggy	85.00
	Total	$924.00

In addition to the above, miscellaneous equipment will be needed which will make the total over $1,000. If not protected from the weather, this equipment would not do good work for more than five years. If well housed, every tool ought to last 12 years or longer. It is obvious that a great saving will accrue by the housing of the implements. An implement house which will house these implements can be built for approximately $200, and it is to be seen that it would prove to be a very good investment.

Sentiment ought to be such that the man who does not take good care of his machinery will be placed in the same class as the man who does not take good care of his live stock.

394. Repairing.—Repairs should be made systematically, and, as far as possible, at times when work is not rushing. It is necessary to have some system in looking after the machines in order that when a machine is to be used it will be ready and in good repair. In putting a machine away after a season's work, it is suggested that a note be made of the repairs needed. These notes may be written on tags and attached to the machine. During the winter the tool may be taken into the shop, with which every farm should be provided, and the machine put in first-class shape, ready to be used upon short notice. It is often an advantage not only in the choice of time, but also in being able to give the implement agent plenty of time in which to obtain the repairs. Often repairs, such as needed, will have to come from the factory, and plenty of time should be allowed.

395. Painting.—Nothing adds so much to the appearance of a vehicle or implement as the finish. An implement may be in a very good state of repair and still give anything but that impression, by the faded condition of its paint. Paint not only adds to the appearance, but also acts as a preservative to many of the parts, especially if they are made of wood.

As a rule, hand-mixed paints are the best, but there are good brands of ready-mixed paints upon the market, and they are more convenient to use than the colors mixed with oil. It is the practice in factories, where the pieces are not too large, to dip the entire piece in a paint vat.

After the color coat has dried, the piece is striped and dipped in the same way in the varnish. This system is very satisfactory when a good quality of paint is used. It is not possible here to give instructions in regard to painting. It might be mentioned, though, that the surface should in all cases be dry and clean before applying any paint.